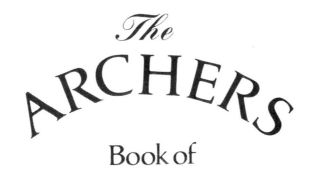

The
ARCHERS

Book of
Farming and the Countryside

ANTHONY PARKIN

BBC BOOKS

To Roberta, with love.

PICTURE CREDITS: COLOUR

Winter Section:
p.97 Hans Reinhard/Bruce Coleman Ltd; p.98 (top) Rare Breeds Survival
Trust; p.98 (bottom) *Farmers Weekly*; p.99 *Farmers Weekly*; p.100 Roger
Wilmshurst/Bruce Coleman Ltd; p.101 (top, centre and bottom) S & O
Mathews; p.102 (top) Massey-Ferguson; p.102 (centre and bottom) S & O
Mathews; p.103–4 S & O Mathews.

Summer Section:
The majority of photographs in this section are by S & O Mathews, with
the following exceptions:
p.172 (bottom) and p.174 (top) *Farmers Weekly*; p.174 (bottom) Derek
Foster; p.175 (top) Patrick Holden; p.175 (bottom) *Farmers Weekly*.

PICTURE CREDITS: BLACK AND WHITE

The majority of the black and white photographs are from *Farmers Weekly*,
with the following exceptions:
p.41 and p.61 Peter Adams; p.72 Massey-Furguson; p.74 *Shooting Times*;
p.77 Peter Adams; p.93 Derek Foster; p.106 Barnaby's Picture Library/
M Campbell Cole; p.109 Derek Foster; p.113 Barnaby's Picture Library/
Edan Fisher; p.119 British Organic Farmers; p.127 Peter Adams; p.130
Derek Foster; p.141 Barnaby's Picture Library/John Woodhouse; p.177
S & O Mathews; p.187 British Organic Farmers; p.200 S & O Mathews;
p.209 and p.213 Peter Adams; p.216 Barnaby's Picture Library/Bill Tasker;
p.226 Peter Adams; p.227 Simon Tupper; p.244 S & O Mathews.

Maps illustrated by Alison Hainey
Cover Illustrations
by Jill Tomblin

Published by BBC Books,
a division of BBC Enterprises Limited,
Woodlands, 80 Wood Lane, London W12 0TT
First published 1989

© Anthony Parkin 1989

ISBN 0 563 20728 0

Set in 11½/13pt Sabon by Phoenix Photosetting, Chatham, Kent
Printed and bound in Great Britain by Mackays of Chatham PLC, Chatham, Kent
Colour separations by Technik Ltd, Berkhamstead
Jacket printed by Belmont Press Ltd, Northampton

Contents

Acknowledgements

Since the early seventies I have 'managed' several farms in Ambridge, ploughing and sowing with the seasons, harvesting and storing the crops, calving, lambing and farrowing. I have introduced new enterprises, taken on fresh labour, updated machinery, bought and sold land, removed hedges and planted trees, all without soiling my hands or risking a bad back.

It has been gratifying to find out how well the farms and the Ambridge countryside have survived their conversion into book form. This is due in large measure to the dozens of people on whom I have leaned over the years in trying to create an authentic and developing agricultural and rural background against which the daily drama of *The Archers* can take place. To each and every one of them I am eternally grateful.

Among those who have been particularly helpful in the writing of this book are John Inge, Jim Vallis, David Richardson, Richie Colwill and Aline and Hugh Black – all of them practical farmers. Patrick Holden and Geoff Mutton helped me considerably with the organic farming, Ian Ballard with the deer and Richard Bach with the Home Farm shoot. I am especially indebted to Don Hardy who prepared the accounts, Tim Angelbeck who read the typescript, Liz Rigbey, lately Editor of *The Archers*, for her encouragement – and to my daughter Sarah who immaculately transposed my rough typing on to her word processor. Others who contributed valuable help included Phil Drabble, Geoff Ballard, and George Jackson of the RASE. Much useful information was provided by the Ministry of Agriculture, Forestry Commission, Milk Marketing Board, Potato Marketing Board, British Sugar, British Wool Marketing Board, National Farmers' Union, Farming and Wildlife Advisory Group and Rare Breeds Survival Trust.

To those and all the others who helped, my thanks.

Anthony Parkin   Tenbury 1989

Ambridge Farms and Farmers

If you wanted to show visitors to this country what makes our farming and country life tick you could do far worse than take them to Ambridge. There they would find a landscape where agriculture and nature coexist in relative comfort, each taking and giving a little as the seasons unfold. It's a productive area, its scenery is rich and varied and it manages to evoke what British farming and the countryside are all about.

You won't find many four-ton an acre crops of wheat there, although Brian Aldridge pulls it off from time to time. There are plenty of farmers with better lambing figures than Phil Archer's and the milk yields at Bridge Farm are never likely to get into the record books. They can't grow early potatoes in Ambridge like they can in Pembrokeshire or corn like they can in Lincolnshire or Brussels sprouts as they do in Bedfordshire. Life is not as hard as it is on the hill farms of Wales or the North or as sophisticated as in the Home Counties (although Jennifer Aldridge may like to think it is). It's quite simply a typical mixture of medium and small farms with the odd large one here and there (like Brian's) to give the other farmers something to envy. Some are owner-occupied, like Phil's, and some like Tony Archer's and the Grundys' are rented. They keep dairy and beef cattle, sheep, pigs, poultry and now red deer, and grow most of the normal crops – wheat, barley, oats, oilseed rape, potatoes, peas, beans and sugar beet – as well as a whole range of vegetables. The countryside, viewed from the top of Lakey Hill, is a rich, intimate blend of hills and valleys, grassland and crops, woods and hedgerows, with the river Am and its tributaries meandering gently between the alders and willows. Those with a keen eye could pick out Brookfield with its collection of old and new buildings, and Phil's barn conversions on the edge of the farm. The tower silo, the only one in the neighbourhood, marks out Home Farm with its huge fields, one of them more than 100 acres. And round to the right lies Tony Archer's organic holding; he would like to think that it *looked* organic but it would be impossible to tell from a distance.

The equilibrium which now seems to exist between farming and nature hasn't come about without stresses and strains. In the sixties and seventies some farmers in the district got too big for their boots, became over-greedy and the environment and wildlife suffered. But nature has a way of reasserting herself, and in the event she was aided and abetted by the very people who had overstepped the mark. Many farmers realised that they had tarnished their image in the eyes of the rest of the community and in many cases damaged their own environment and lowered the value of their own farms. In the last few years people have been planting trees and creating ponds and other habitats and in some cases even re-establishing hedges in an effort to make amends.

Not that things in Ambridge ever went very far in the wrong direction. The countryside thereabouts is just too well-stocked by nature to be destroyed easily; there are so many trees, so many banks and dingles and wet patches too difficult to plough which have to be left alone. The local population is very keen on its field sports and needs the proliferation of woods and copses to hold foxes and pheasants; paradoxically, hunting and shooting are the greatest friends of conservation. And finally there are more Joe Grundys in the neighbourhood than you would expect, with their old-fashioned outlooks who 'don't hold with all these new-fangled techniques'. It was no coincidence that when the otter came to Ambridge a few years ago she settled on Joe's overgrown, untidy stretch of the river Am, where Eddie rather than taking the trouble to burn the lop and top from some fallen trees had simply pushed them into the river with his tractor.

It would be surprising if Ambridge had escaped totally unscathed from the agricultural revolution which has brought more change to our farms over the last 40 years than at any similar period in history. But an old Ambridge countryman, returning to his birthplace for the first time since the war, would certainly not find the prospect displeasing. What would strike him? One of the first things, perhaps, would be that there was hardly anyone in the fields to talk to. There are only about a quarter the number of farm workers that there were when he left Ambridge and any he did encounter would probably be cocooned inside a tractor cab. He'd notice that all the Shorthorn cattle had gone from the fields to be replaced by black and white Friesians or one of the beefy Continental breeds imported over the last 25 years. He might look at the empty stands at the ends of farm drives and wonder where the milk churns had gone – until he had to step into the ditch as the tanker thundered past. Hedges trimmed rather than laid, bigger fields, modern buildings – he'd note it all, and be amazed at how little it had changed.

Those, like Phil Archer, who have been in Ambridge all their lives, would know that things had altered considerably. He has seen some farmers go out of business while others prospered. Brookfield itself has grown, from the 100 acres it was when the fresh-faced Phil left the Farm Institute, to its present 462 acres. Willow Farm, the small dairy farm which gave Tony Archer a start on his own in the seventies, is no longer a viable holding. Ironically, it was Phil who bought it in 1983 and 'asset-stripped' it, selling off 30 acres to Brian Aldridge, the house and 15 acres to Bill Insley and keeping the remaining 55 acres for himself at a very favourable price. Since then the property has been split even further, with Neil Carter acquiring a share and the house being bought by the local doctor as an investment; very much a sign of the times.

Another sign of the times was the bankruptcy of Mike Tucker in 1986. Mike's trouble was that he began farming on his own too late and with too little money. His was the classic case of the under-capitalised young man entering farming at the tail-end of the boom period, to whom the bank had lent far too much money. He didn't own Ambridge Farm so he had no collateral when the bank wanted to foreclose. Now Dr Thorogood lives in his old farmhouse and the land (including the milk quota) is farmed by the Bellamy estate.

Farming has always been subject to alternating periods of prosperity and depression. The downturn of the 1980s comes at the end of an unparalleled boom of 40 years, virtually the whole of Phil Archer's working life. The irony is that it has been brought about by agriculture's own success as Ambridge farmers, like their fellow producers this country and across the Channel, have been turning out more milk and bacon, more wheat and barley than we can eat.

Here are some of the changes of the last 40 years:

	THEN	NOW
Number of farms (UK)	527 000	144 000 (viable)
Average farm size (UK)	63 acres	260 acres (viable)
Regular farm workers (UK)	714 000	200 000
Price of land/acre (E)	£50	£1500
Working horses (GB)	350 000	virtually none
Nitrogen (plant food)	120 000 tonnes	1 200 000 tonnes
Wheat yield/acre	1.1 tonnes	2.5 tonnes
Dairy herds (E & W)	160 000	36 000
Milk/cow/year (E & W)	620 gallons	1100 gallons
Average dairy herd (E & W)	16 cows	62 cows
Eggs/hen/year	150	262
Poultry meat	100 000 tonnes	1 000 000 tonnes

Methods of collecting and presenting statistics have changed considerably over the years making it difficult to be precise, but these figures represent the general trends. It has not been possible to base them all on United Kingdom (UK) figures. Some are statistics for Great Britain (GB), some England and Wales (E & W) and some for England only (E).

If Ambridge is a microcosm of British agriculture, the farms of Phil Archer, Brian Aldridge and Tony Archer are typical of thousands up and down the country: Brookfield the classic mixed family farm run by father and son, Home Farm the large arable farm which provides its owner with a flamboyant lifestyle, and Bridge Farm the small tenanted holding where life has always been something of a struggle.

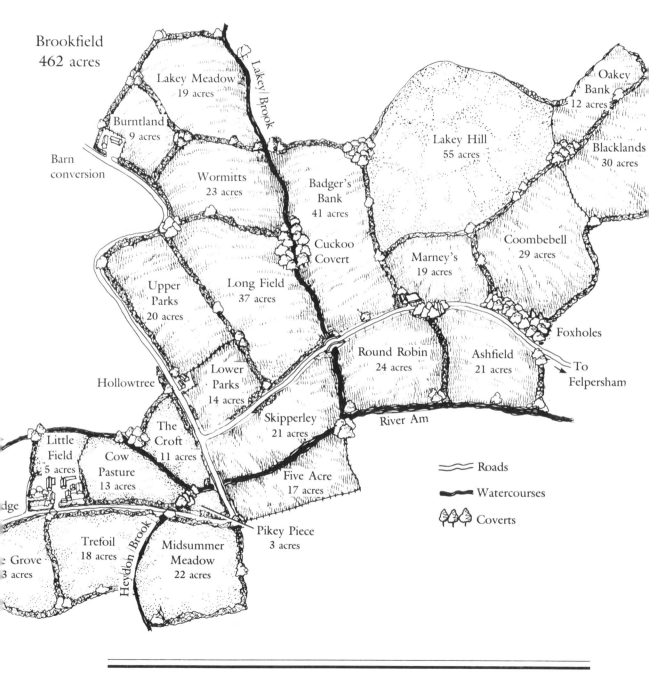

Brookfield
462 acres

Lakey Meadow
19 acres

Lakey Brook

Burntland
9 acres

Barn
conversion

Wormitts
23 acres

Badger's
Bank
41 acres

Lakey Hill
55 acres

Oakey
Bank
12 acres

Blacklands
30 acres

Cuckoo
Covert

Coombebell
29 acres

Upper
Parks
20 acres

Long Field
37 acres

Marney's
19 acres

Foxholes

Hollowtree

Lower
Parks
14 acres

Round Robin
24 acres

Ashfield
21 acres

To
Felpersham

The
Croft
11 acres

Skipperley
21 acres

River Am

Little
Field
5 acres

Cow
Pasture
13 acres

Five Acre
17 acres

Roads

Watercourses

Coverts

dge

e Grove
3 acres

Trefoil
18 acres

Heydon Brook

Midsummer
Meadow
22 acres

Pikey Piece
3 acres

BROOKFIELD

Brookfield is a real mixed family farm. Phil and David, now in partnership, are both actively involved in the business, and Jill is a typical supportive farmer's wife. Unlike many farms they still keep cattle, sheep and pigs and grow a wide range of crops. Dan Archer resisted the post-war encouragement to specialise, possessing a lifelong fear of putting too many eggs into one basket, a philosophy which Phil seems to have inherited. Although the dairy herd has been the centre of the farm over the years it has been well-supported by the other livestock and arable enterprises and when one sector has been in the doldrums another has usually been thriving.

The somewhat elongated shape of the holding, with the farmhouse and buildings situated at one end, is explained by the acquisitions of the last 30 years. When Phil was a boy Brookfield was 100 acres, rented from the Lawson-Hope estate. Now it extends to 462 acres and is, to all intents and purposes, owner-occupied. Dan Archer bought the farm from his landlord when the estate was sold in 1954, but the main expansion took place in the early sixties when he got together with his two neighbours, Jess Allard and Fred Barratt, to form Ambridge Farmers Ltd. By the end of that decade Dan had acquired their two farms, bringing Brookfield up to 435 acres. Then followed a period of consolidation until 1979 when another neighbouring holding, Meadow Farm, came on to the market. Few farmers can resist the opportunity to acquire adjacent land – 'we'll never get the chance again' is the usual justification – and Phil, by then in charge, bought a further 30 acres. Four years later he purchased Willow Farm at the other end of his holding and retained 55 acres, selling the remainder to Brian Aldridge and Bill Insley. This proved a short-lived addition since, following the death of his father, he was forced to resell this parcel to meet a tax bill. He parted with a further 3 acres when he sold the barn and adjoining buildings for conversion into houses, leaving the present total.

Milk quotas, introduced in 1984, forced Phil to cut his dairy herd from 110 Friesians to 95, although he and David are now considering leasing some extra quota since they have the buildings, milking parlour and labour to cope with more cows. This is one of the options as they consider how to face up to a future of falling margins on most of the things they produce. However, Brookfield faces fewer pressures than many similar-sized farms since the bulk of the acreage was purchased at a fraction of its present day value.

The land is basically fertile although rather on the heavy side. This can cause irritation in the spring as Phil sees his neighbour, Brian Aldridge, busy ploughing or cultivating on his lighter land while he is forced to wait until his land is drier and more workable. It grows good grass, which accounts for more than half the acreage, and produces quite respectable yields of cereals and root crops. Parts of the farm are too steep to plough.

Like most farmers these days Phil likes to keep the labour force to a minimum, bringing in casual workers or contractors to help out when necessary. He still does quite a lot of physical work himself and can tackle virtually any job on the farm. David is probably happier in the driving seat of a tractor than mucking out calf pens, but is quite capable of milking, trimming sheep's feet or injecting piglets when necessary. Graham Collard copes well with the dairy herd and sometimes offers himself for other work at busy times, while Bert Fry has shown himself to be a dependable all-rounder. Neil is employed on a part-time basis to run the pig unit at Hollowtree and is sometimes roped in to help with silage, haymaking or harvest. With Ruth available at times, they are well-staffed.

---

462 ACRES

STOCKING:	CROPPING: average acreage*		LABOUR:
Dairy cows 95	Cereals	140	Phil Archer
Followers 75 (heifers and calves)	Oilseed rape	35	David Archer
Beef cattle 50 (various ages)	Beans	12	Graham Collard (cowman)
Sows 60 (progeny sold as baconers)	Potatoes	15	Bert Fry (general)
Ewes 300	Fodder beet	7	Neil Carter (pigs, part-time)
	Grassland	250	Ruth Archer (when available)

acreages vary from year to year according to field sizes

Brookfield

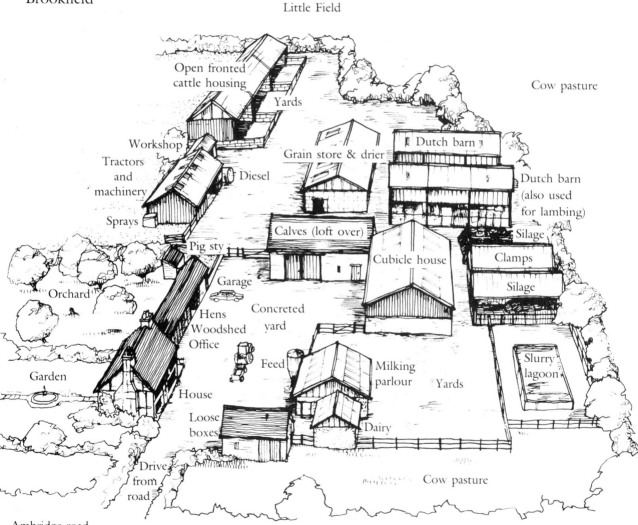

Little Field

Cow pasture

Open fronted
cattle housing

Yards

Dutch barn

Workshop

Grain store & drier

Dutch barn
(also used
for lambing)

Tractors
and
machinery

Diesel

Sprays

Calves (loft over)

Silage

Pig sty

Cubicle house

Clamps

Garage

Silage

Orchard

Concreted
yard

Hens
Woodshed
Office

Slurry
lagoon

Feed

Garden

Milking
parlour

Yards

House

Loose
boxes

Dairy

Drive
from
road

Cow pasture

Ambridge road

PROFILE OF PHIL ARCHER

There's a Phil Archer in most rural communities, a respected farmer who has, in many cases, come to the fore as a direct result of forty years of agricultural prosperity coupled with the decline of the local squirearchy. In Ambridge, Jack Woolley may be richer and Brian Aldridge may have a bigger farm but it's Phil Archer whom the villagers have come to trust. The Archers are synonymous with Brookfield, having farmed there for more than a century, which further enhances their status.

Typical of many farmers of his generation he's hard-working and conscientious but shrewd and careful, mild-mannered but determined, quiet but capable of anger; he presides over his family and farm in a relaxed but decisive way. His wife and children find him infuriating at times, which doesn't stop them from loving him, and he's respected by most of his friends, neighbours and workers.

Phil's a good farmer but not an innovator. He has never forgotten his father's old adage 'Be not the first by whom the new is tried – nor yet the last to cast the old aside'. So he let his neighbour, Brian Aldridge, try out oilseed rape first; only when he had watched him overcome the teething troubles of a new crop for a few years did he establish it as part of the Brookfield rotation. He tries to keep his son David's exuberance in check, remembering how similar he was at the same age.

By nature he's no zealot. He's a good Christian but not fanatical and would happily take the combine out on a Sunday to win a crop of wheat or barley in catchy weather. In his day-to-day life he's rather non-political although in the privacy of the ballot box he almost certainly votes Conservative. He's a paid-up member of the National Farmers' Union and dutifully did his year as branch chairman, but was happy to relinquish the post when the time came.

He plays his part in local affairs and has been a magistrate on the Borchester bench for years. Nearer home he's a member of the Parochial Church Council and has been known to umpire cricket matches from time to time. His wife, Jill, is a member of the Ambridge Women's Institute and is closely involved with the Women's Royal Voluntary Service, including playing an active part in the local Meals on Wheels operation.

Running a large mixed farm and discharging his family and other responsibilities leaves Phil little time for hobbies. His main relaxation is listening to music and playing the organ in church. David would argue that his obsession with Freda, the Middle White pig, and her offspring is a hobby although Phil would no doubt dispute this. Not a great drinker, he enjoys a glass of sherry or the odd half-pint of beer or occasional whisky. Neither is he a great one for holidays. Most of his trips, like his recent NFU visit to Australia, have something to do with farming.

Phil and Jill live comfortably but relatively modestly. They want for nothing that really matters but indulge in few luxuries. The 'profit' from Brookfield, when eroded by tax, repayment of bank loans, provision of a pension and reinvestment in the farm leaves little enough for a high lifestyle even if they wanted it, but the perks of the job ensure that they live well. Phil is content to drive a three-year-old Sierra, while Jill gets by in an even older hatchback. Only David makes a bit of a splash with his XR3i.

ANNUAL ACCOUNTS			
RECEIPTS		**COSTS**	
	£		£
Milk	84 000	Livestock	11 000
Cattle	23 000	Feed	89 000
Lambs	16 000	Seed	7000
Sheep	5000	Fertiliser	20 000
Pigs	81 400	Sprays	9000
Cereals	36 600	Labour	35 000
Rape	9800	Machinery, contract work, repairs, fuel, etc	22 000
Beans	3200	Machinery depreciation	21 000
Potatoes	15 000	Finance charges	14 200
Miscellaneous (inc. produce consumed)	6000	Others	12 800
Total output	£280 000	Total costs	£241 000

'Profit' = £39 000 = £84 per acre

BALANCE SHEET

Land: £693 000

OTHER ASSETS		LIABILITIES	
	£		£
Machinery	92 000	Creditors	42 000
Livestock	111 500	Bank overdraft	90 000
Crops, stores, etc	47 000		
Debtors	11 000	AMC mortgage*	40 000
Cash at bank	1 500		
	263 000		172 000

Phil Archer's net worth (including land) = £784 000

*The AMC (Agricultural Mortgage Corporation) mortgage is in respect of the land he bought when Meadow Farm was split up in 1979.

GUIDE TO FARM ACCOUNTS

The 'profit' from a farm is not easily defined and very difficult to relate to a wage or salary.

The figures shown here are the return the farmer receives for his own labour and management and capital invested in the farm. Out of it must come any tax, repayment of bank loans and future investment over and above the depreciation figure already allowed for in the accounts. (For example, at Brookfield, £21 000 has been allowed for machinery depreciation. Should Phil Archer need to spend more than that, the balance would have to come out of 'profit'. Because machinery is constantly getting dearer and because they are developing their businesses most farmers are increasing their investment all the time. The tax is based not on the 'profit' alone but on the sum of the 'profit' and the machinery depreciation; in Brookfield's case this would be £60 000.)

To the 'profit' figures must be added the perks of the job – milk, meat, eggs, vegetables, firewood, plus housing, telephone, heating, car expenses, etc, most of which come 'out of the farm' (though the tax inspector would expect a say in what proportion of these are farmhouse as opposed to farm expenses). People running small businesses will find the situation easier to appreciate than those earning salaries. The difference between farming and other businesses is that a farm provides what many people see as an enviable way of life as a bonus. The owner of an engineering works, for example, can't eat his own products, or keep a horse or two at the factory; and if he wants a country life he must travel to work. On the other hand, 'living over the shop', often in relative isolation from shops, schools and other facilities, might be regarded by some as a disadvantage.

At first sight, both Brian and Phil appear to be doing quite well, but if the land is taken out of the equation the situation is far less rosy. If Brian Aldridge were to charge himself a notional rent (say at the same level as Tony Archer is paying – £50 an acre) he would find that he was actually making a trading loss. Although he is able to maintain a lavish lifestyle he is, in fact, living off his investment in the land. If the same were applied to Phil Archer, his profit would be cut to £11 000 before tax. Only Tony Archer's is a genuine trading 'profit' and it hardly pays for the physical effort he and his wife put into the farm, particularly during the transition to an organic system.

It's one of the features of farming that, because so many of the daily necessities of life come out of the business, a farmer can appear to enjoy a much better lifestyle than the 'profit' would justify. He might hunt three days a week, have the horse fed for 'nothing', the Land Rover and trailer run largely by the business and pay no hunt subscription because he is a farmer – and all while the accounts were showing a loss.

The accounts shown in these pages are the most recent ones available but pre-date the setting up of the Brookfield partnership and the Bridge Farm move into organic yoghurt.

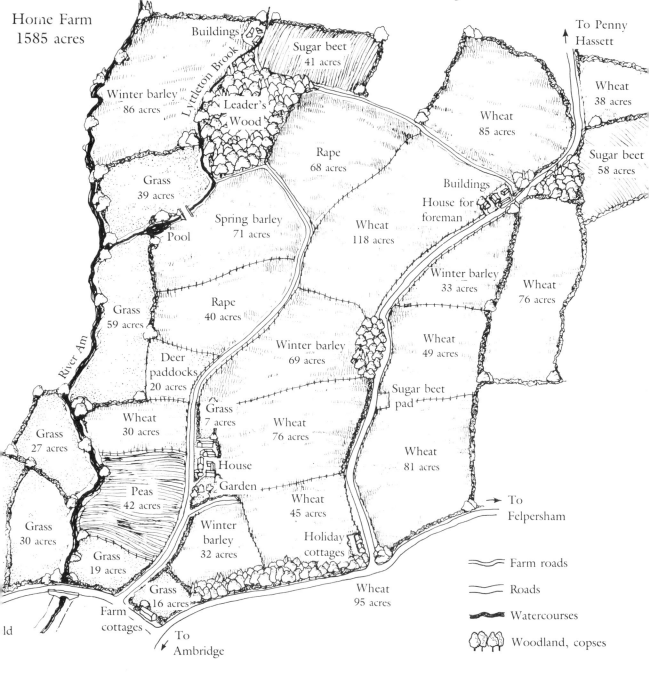

Home Farm
1585 acres

Buildings

To Penny
Hassett

Sugar beet
41 acres

Winter barley
86 acres

Lyttleton Brook

Leader's
Wood

Wheat
38 acres

Wheat
85 acres

Rape
68 acres

Sugar beet
58 acres

Grass
39 acres

Buildings

House for
foreman

Spring barley
71 acres

Pool

Wheat
118 acres

River Am

Grass
59 acres

Rape
40 acres

Winter barley
33 acres

Wheat
76 acres

Deer
paddocks
20 acres

Winter barley
69 acres

Wheat
49 acres

Grass
7 acres

Sugar beet
pad

Wheat
30 acres

Grass
27 acres

Wheat
76 acres

Wheat
81 acres

House

Garden

Peas
42 acres

Wheat
45 acres

Grass
30 acres

Holiday
cottages

Winter
barley
32 acres

Grass
19 acres

To
Felpersham

Grass
16 acres

Wheat
95 acres

Farm
cottages

ld

To
Ambridge

〰〰 Farm roads

── Roads

〰 Watercourses

🌳 Woodland, copses

HOME FARM

By the time Brian Aldridge acquired it in 1975 Home Farm was an amalgamation of several holdings or parts of holdings. At 1500 acres – now 1585 with the addition of two parcels of Willow Farm land – it is much bigger than the average for the area. For many years the eighteenth-century farmhouse was let as flats and the land farmed as part of Ralph Bellamy's large estate. It was Bellamy who began ripping out the hedges to make the fields bigger and more suitable for large machinery but Brian has certainly played his part, with the result that one of his fields is now more than 100 acres, and the average is over 60 acres (half the size of Joe Grundy's whole farm).

More than half the fields can be reached from a council road – always a great advantage to an arable farmer – and Brian has put in farm roads to give access to most of the others to carry the thousands of tons of crops which must be transported each year. The land tends to be a little lighter than elsewhere in Ambridge, making it ideal for the cropping pattern Brian has adopted. Nearly all of it is ploughable, although so far Brian has not broken up the grass fields on the Brookfield side of the farm. His aim is to re-seed some of the arable fields in rotation so that they can be grazed by his large sheep flock and contribute the fertility which comes from 'the golden hoof'.

Brian came to Ambridge from a much smaller farm in Hertfordshire; this was sold for development which provided him with enough money to purchase Home Farm. To the extent that he doesn't have a large mortgage to pay off or a big rent to find twice a year he has an inbuilt advantage when times get difficult. But no farmer, least of all Brian, likes to see his income falling, and he can be expected to continue his search for alternative ways of making money from the land.

As profits from cereal growing fall, he puts a lot of effort into growing quality grain – seed crops, milling wheat, malting barley – all of which earn him a valuable premium over corn destined for animal feed. Worried about oilseed rape, he is trying out peas as an alternative. On the livestock side, the sheep, lambed in two batches to ease demands on labour, have done well and fit in with the rest of the farm. The beef cattle do not make as much money, but Brian likes to see them around and they utilise some of the arable by-products such as straw and sugar beet tops. He plans to continue building up the deer enterprise as long as he can see a profitable market for the hinds and some of the stags as breeding stock.

There is a minimum staff of well-paid workers at Home Farm, augmented by students at busy times such as lambing and harvest. The farm is very well-mechanised with six tractors – including a 130 hp four-wheel-drive monster – and three combine harvesters. The total cost of replacing all the machinery would be more than half a million pounds.

1585 ACRES

STOCKING:	CROPPING: average acreage*		LABOUR:
Ewes – 600	Cereals	1010	Brian (managing and relief)
Deer 85 hinds (plus followers)	Oilseed rape	120	Working foreman
Beef cattle 100 (bullocks fattened/year)	Peas	40	Sammy Whipple (stockman)
	Sugar beet	100	3 tractor drivers
	Grassland	220	1 boy
	Woodland	80	

actual acreages vary from year to year according to field sizes

---

Home Farm

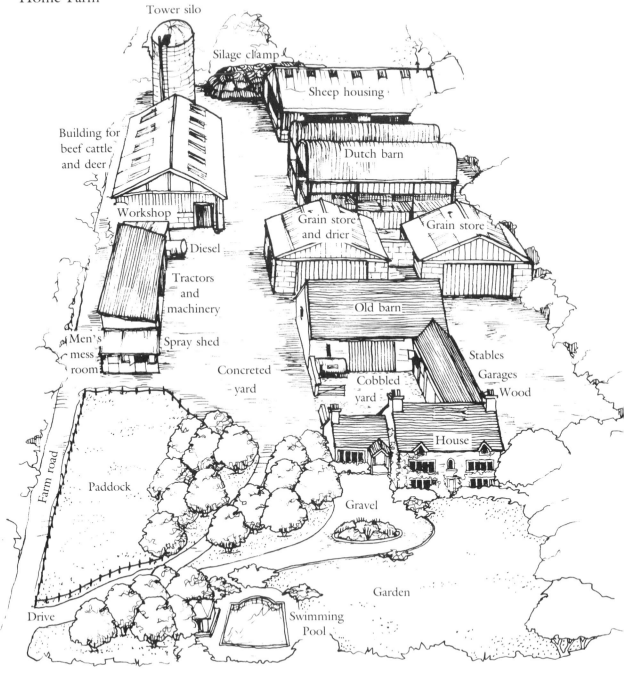

Tower silo

Silage clamp

Sheep housing

Building for
beef cattle
and deer

Dutch barn

Workshop

Grain store
and drier

Grain store

Diesel

Tractors
and
machinery

Old barn

Men's
mess
room

Spray shed

Stables

Garages

Concreted
- yard

Cobbled
yard

Wood

House

Farm road

Paddock

Gravel

Garden

Drive

Swimming
Pool

PROFILE OF BRIAN ALDRIDGE

Brian is a product of the boom years of British agriculture. Born at the end of the last war he has, until the last few years, lived in a world of sound price guarantees, of generous grants and subsidies and of continuing technological improvement. It was a period when farmers were constantly being exhorted to increase production, when the Ministry of Agriculture peddled unlimited free technical advice and when the banks were queueing up to lend them money. Things like quotas and co-responsibility levies were things of the future. It has been said that only a fool could fail to succeed in farming during this time and Brian Aldridge is certainly no fool. As if all this and a public school education were not sufficient advantage, his holding in Hertfordshire was sold for enough money to enable him to buy Home Farm, Ambridge, without a mortgage.

Although he would boast, at least until his recent accident, that he's capable of doing any job on the farm he has no intention of doing most of them. He's an accomplished delegator and has built up a trustworthy team to support him. While they plough and sow and dip sheep and dig out blocked drains, Brian is in his office – on the telephone to dealers, brokers and bankers, and operating his computer. He's quite capable of lambing a ewe or driving the combine if absolutely necessary – but as he tells Jenny, his wife, that's not what he keeps a staff for.

His 'voluntary work' tends to be of the sort which offers reciprocal advantages in terms of productive information or meeting useful people. He's a member of the NFU and the Country Landowners' Association and the Oxford Farming Conference. He's a Governor of the Royal Agricultural Society of England, a Fellow of the National Institute of Agricultural Botany, on the Board of Borsetshire College of Agriculture and a director of Bunter McColl, the machinery distributors.

Although he has probably destroyed more hedges, uprooted more trees and filled in more ponds than any other farmer in Ambridge he was, paradoxically, a founder member of the local Farming and Wildlife Advisory Group. Like many fellow farmers who abused the countryside to line their pockets during the sixties and seventies, he has been in the van of those trying to put things right in the eighties and has been digging a small lake and planting trees.

Brian certainly knows how to enjoy himself when he's not farming. He's a lover of good food and wine and likes eating out. He has friends all over the country and thinks nothing of driving up to Scotland for a day's shooting and an evening with an old chum, or over to East Anglia to look round a farm. His holidays tend to be typically upmarket, although curiously he's not as keen to get away as might be expected. If it's sun he's after it has to be somewhere like the West Indies or the Seychelles, while his usual skiing resort is Kitzbuhl.

His favourite vehicle seems to be his off-white Range Rover which is seen regularly both on and off the farm, with Brian clad in roll-neck jersey and Barbour or shirt and red-and-white spotted cravat — often speaking on the carphone or short wave radio. There's a last year's BMW in the garage for the longer journeys, alongside his wife's Volvo estate.

ANNUAL ACCOUNTS

RECEIPTS	£	COSTS	£
Cereals	307 700	Sheep	11 400
Rape	35 000	Cattle	48 400
Sugar beet	60 000	Feed	7000
Peas	7300	Seed	27 300
Sheep	46 600	Fertiliser	58 400
Deer	10 800	Sprays	53 700
Cattle	58 800	Labour	70 000
Miscellaneous (inc. produce consumed)	12 000	Machinery, contract work, fuel, etc	60 000
		Machinery depreciation	71 000
		Finance charges	24 000
		Others	36 000
Total output	£538 200	Total costs	£467 200

'Profit' = £71 000 = £45 per acre

BALANCE SHEET

Land: £2 220 000

OTHER ASSETS		LIABILITIES	
	£		£
Machinery	320 000	Creditors	38 000
Livestock	100 000	Bank overdraft	95 000
Crops, stores, etc	143 000	Bank loans*	115 000
Debtors	12 000		
Cash at bank	3000		
Total assets	£578 000	Total liabilities	£248 000

Brian Aldridge's net worth (including land) = £2 550 000

His bank loans are for the purchase of two parcels of Willow Farm land.

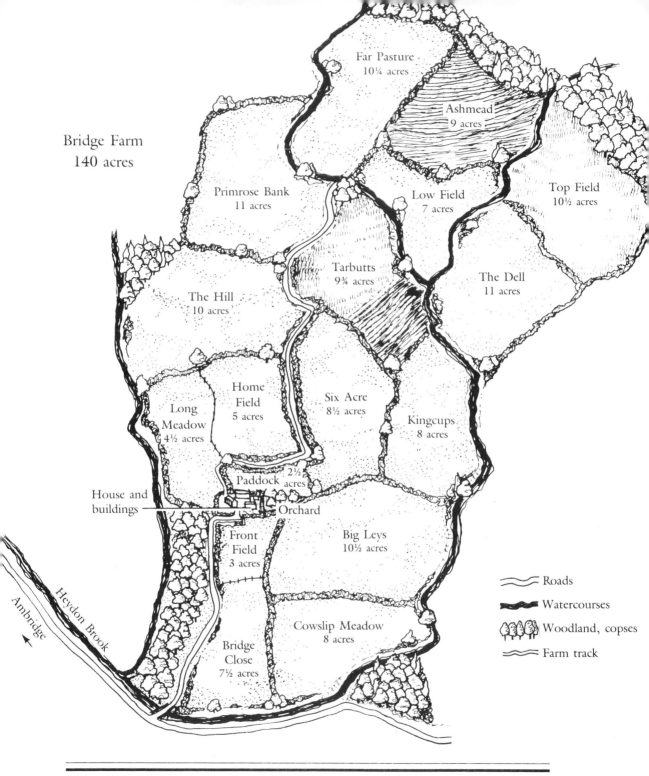

Bridge Farm
140 acres

Far Pasture
10¼ acres

Ashmead
9 acres

Primrose Bank
11 acres

Low Field
7 acres

Top Field
10½ acres

Tarbutts
9¾ acres

The Hill
10 acres

The Dell
11 acres

Home
Field
5 acres

Six Acre
8½ acres

Kingcups
8 acres

Long
Meadow
4½ acres

Paddock
2½ acres

House and
buildings

Orchard

Front
Field
3 acres

Big Leys
10½ acres

Heydon Brook

Ambridge

Cowslip Meadow
8 acres

Bridge
Close
7½ acres

〰 Roads

〰 Watercourses

🌳 Woodland, copses

〰 Farm track

BRIDGE FARM

Life for Tony and Pat Archer changed dramatically in 1984, the year they decided to 'go organic'. Until then, as on many smaller dairy farms, the aim had been to milk as many cows as possible. They had already built up the herd to 100 Friesians and planned to go even further when suddenly milk quotas were imposed. This would have meant an immediate reduction in cow numbers to avoid producing more than their quota, but they decided to go much further.

Uneasy at what they were doing to the land, the cows and themselves in plastering their grass with fertilisers to supply milk which no one wanted, they made up their minds to cut the herd by about a third and try to manage without chemicals altogether. The plan was to convert their farm to the organic system over a five-year period, taking in 30 acres a year. By bringing in the last two years' acreages in one go they managed to finish the conversion in 1988 and can now use the Soil Association symbol for produce from the whole farm. Simply to stop using nitrogen on the grassland and routine antibiotic treatment of the cows would have been relatively simple but, in order to keep up the cash flow during the transition, they had to start growing a range of new crops, cashing in on the premium paid for organic produce.

Bridge Farm is an ideal dairy holding with a pattern of small fields, most of which are watered by streams. A track running up through the middle of the farm means that most of the fields can be reached easily. In addition to the main holding Tony rents a further 10 acres of land near the village which he uses either to make hay or to graze young stock.

The farm is rented from the Bellamy estate, 1000 acres of which were retained after it was split up following the death of Ralph Bellamy. Tony is now in the unenviable position of having his sister Lilian as his landlord, and his cousin Shula Hebden acting on her behalf through the agents Rodway and Watson. He has seen his rent rise steadily during the 11 years he has been a tenant; at the last triennial review in 1987 it was rounded up from £48 to £50 an acre. With farming going through something of a depression Tony would expect the rent to be held at that level when it next comes up for review in 1990.

After several years of selling through various wholesalers, Tony and

Pat have now signed a contract with a co-operative specialising in organic produce. This binds them to sell at least 90 per cent of their vegetables in this way, leaving them free to market the rest locally. Wheat is not covered by the agreement and they have some of it ground for sale as organic flour but, as the acreage increases, most of it goes direct to a grain merchant, again at a substantial premium over conventional wheat. The most recent development is the move into yoghurt and other dairy produce; this has had a significant effect on the return from their milk which, otherwise, goes to the Milk Marketing Board without any premium for being organically produced.

With only Tony, Pat and Steve, the YTS lad, on the staff (the latter two both part-time), the farm is permanently under-manned; casual workers are brought in to cope with planting, hoeing, harvesting and packing the crops when Tony and Pat cannot manage, and contractors deal with silage making and combining.

---

140 ACRES (+ 10 acres grazing near village)

STOCKING:	CROPPING: average acreage*		LABOUR:
Dairy cows 65	Spring wheat	15	Tony Archer
Followers 50 (heifers and calves)	Potatoes	6	Pat Archer
Beef cattle 10 (variable)	Leeks	4	Steve (YTS)
	Carrots	4	Casual labour for hoeing, harvesting crops
	Swedes	3	
	Dutch cabbage	2	Contractors for silage-making, combining
	Savoy cabbage	1	
	Grassland	115	

acreages vary from year to year according to field sizes

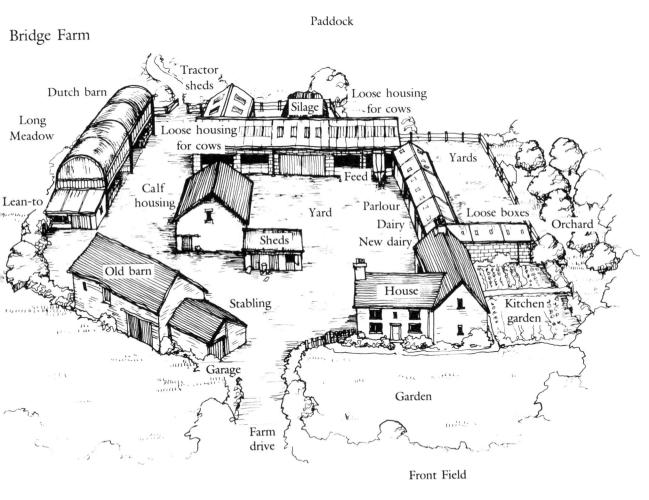

Bridge Farm

Paddock

Dutch barn

Tractor sheds

Long Meadow

Silage

Loose housing for cows

Loose housing for cows

Lean-to

Yards

Calf housing

Feed

Yard

Parlour
Dairy
New dairy

Loose boxes

Orchard

Sheds

Old barn

House

Stabling

Kitchen garden

Garage

Garden

Farm drive

Front Field

PROFILE OF TONY ARCHER

Although not a farmer's son Tony was determined to get into farming, where his grandfather Dan and Uncle Phil were already firmly established. After attending Shropshire Farm Institute and working on the Bellamy estate his chance to set up on his own came in 1973 when he was offered a 'landlord-tenant partnership' on Willow Farm. The landlord (Haydn Evans) provided the land and much of the capital and Tony was to farm it. Five years later, now married to Pat, he obtained the tenancy of Bridge Farm, his present holding.

He and his wife went through a difficult phase during the early eighties from which they now seem to have emerged with the partnership much stronger. Going organic brought them closer together both philosophically, with a growing belief in the soundness of what they had embarked on, and physically because their new way of farming meant much more work on the farm for both of them. As a result Tony has become a somewhat less frequent visitor to The Bull (and less inclined to follow his alcoholic father) and Pat has tempered her forays into the peace movement and feminist activities which at one time threatened to engulf her. They have discovered that making a success of a complex organic farm with limited labour needs their full-time combined efforts.

It's a sadness to Tony – as it would be to any farmer – that his elder son prefers American football to farming and is disinclined to help unless heavily bribed. However, the younger one, Tommy, is shaping up nicely and follows his father everywhere when given the chance.

Tony, at 39, still suffers from having a chip on his shoulder as he tries to come to terms with the fact that one of his sisters owns the land he farms while the other is married to the richest farmer in South Borsetshire. His cousin David has just been made a partner in the family-owned farm of nearly 500 acres, and still Tony struggles to make enough to meet the twice yearly rent demand.

However, since the move into making yoghurt and other dairy products, prospects at Bridge Farm are set to improve and he has the satisfaction of knowing that, while his brother-in-law's and cousin's farms are still producing commodities which are in surplus, there is an ever-growing demand for his organic produce.

Tony is a typically loyal but inactive member of the NFU and spends more time these days with a local group of organic farmers and growers. He's also on the committee of the Borchester Agricultural Society through his involvement in showing beef cattle. Pat keeps up her connection with the Borchester Women's Group and the Film Society at Borchester Tech, though far less aggressively than at one time. In fact, rather against her principles, she has recently become a supporter of the local Pony Club in response to daughter Helen's enthusiasm.

The W reg Ital and the old Land Rover are due for updating if the yoghurt-making is successful in the long run.

ANNUAL ACCOUNTS			
RECEIPTS		**COSTS**	
	£		£
Milk	52 500	Livestock	1000
Livestock	12 400	Feed	26 700
Wheat (inc. flour)	4500	Packaging, etc	5200
Potatoes	4300	Seeds and plants	3000
Carrots	3600	Organic fertilisers	1200
Leeks	6000	Labour	10 400
Swedes	3100	Power and machinery (inc. contract)	15 200
Dutch cabbage	1800	Rent and rates	8000
Savoys	1200	Finance	3400
Miscellaneous (inc. produce consumed	4700	Others	5000
Total output	£94 100	Total costs	£79 100

'Profit' = £15 000 = £100 per acre

BALANCE SHEET

ASSETS		LIABILITIES	
	£		£
Buildings	4000	Creditors	10 000
Machinery	18 000	Bank overdraft	20 000
Livestock	51 100	Bank loan	10 000
Crops, stores, etc	3200		
Debtors	8500		
Cash at bank	200		
Total assets	£85 000	Total liabilities	£40 000

Tony Archer's net worth = £45 000

October

There is no clear-cut beginning or end to the farming year; no date on which all of the preceding season's crops have been harvested and none of the following sown. Custom prescribes that October is the start of the new year and it would be difficult to find a more natural date, though there are lots of ifs and buts.

True, the corn harvest is over and the combines put back under cover for another year. But Brian hasn't started lifting his sugar beet, Phil's potatoes are still in the ground and most of Tony's organic crops are waiting to be turned into cash. On the other hand, while the bulk of the winter wheat and barley has yet to be drilled to ripen next summer, some of it is already in the ground and showing through. And the oilseed rape, the crop which splashes the countryside with vivid yellow each May, will have been planted as early as the end of August, only a few weeks after the previous crop was harvested.

October is a 'comfortable' month in the country. The weather is usually reasonably kind and often includes an Indian summer about half-way through, or St Luke's little summer as it's sometimes known (the saint's day is on 18 October). The skies can be clear blue as the butterflies settle on the Michaelmas daisies in the garden of Glebe Cottage. Tom Forrest knows where there are still a few mushrooms at the beginning of the month and Prue may be picking blackberries until a touch of frost puts an end to these gifts of nature. Ambridge gardens are bulging with vegetables; the potatoes and carrots are ready to be lifted and stored while the cabbages and sprouts will be left to face the frost. There are apples to be picked and chutney to be made, damson and elderberry wine to be started and, later in the month perhaps, the more affluent will be making a bottle or two of sloe gin for Christmas and after.

Ambridge looks at its most attractive in October with trees of every sort blending in colour from green through brown, yellow and gold to red as the leaves turn. The last of the swallows gather on the wires near the village green before starting their long flight to winter in Africa. Meanwhile, the Horobin youngsters are in danger of breaking the windows of The Bull as they pitch sticks and stones at the chestnut tree outside to bring down the conkers – until Sid Perks comes out to drive them away again.

Hunting runs through the social life of the countryside, and although the opening meet of the foxhounds isn't usually held until the first Saturday of November the hunt will have been out cubbing since the last of the corn was cut in mid-September. Pheasant shooting begins officially on 1 October but most shooting folk hold off for a few weeks to allow the young birds to get bigger. But there are more partridge around these days since farmers like Brian Aldridge have seen the error of their ways and begun to leave an unsprayed strip around their fields. Partridge shooting began on 1 September and Brian and his chums will certainly be having a go at them during October. And some time this month Phil, Tony, Joe and other local farmers will probably receive a card from the secretary of the local pack of beagles asking for permission to look for a hare on their land. They're not likely to refuse; it's not thought very sporting to say no.

Early in the month the women in the parish will be gathering at St Stephen's to decorate the church for Harvest Festival and someone will be pressed into organising yet another harvest supper. Towards the end of the month Phil – if he has time – may enter some carefully washed potatoes or fodder beet or perhaps some barley or a sample of silage for the Borchester Agricultural Society's Autumn Show.

BROOKFIELD

Phil's first main job is to put the rams in with the ewes, which he likes to do as nearly as possible on the first day of the month. This means that his flock will start to lamb on 25 February as the gestation period is 147 days. There's nothing very spectacular about putting the tups in (tup is another name for ram). Phil divides his flock of 300 ewes into two bunches and puts four rams in with each; they usually set to work straight away. To prepare the ewes for this he will have been keeping them on some good pasture for the previous month. This is called 'flushing'. The idea is to have the ewes on a rising plane of nutrition which encourages them to ovulate more freely and so increase the potential number of lambs born. The only preparation the ram undergoes is to have some raddle (a coloured powder mixed with oil) smeared on its chest, between its front legs. When it mounts a ewe to mate, some of this rubs off on the ewe's hindquarters and remains visible for several weeks. This makes it easy to keep a check on which ewes have received a visit. By using different coloured powders as the season progresses Phil can tell which ewes have been mated and in what order, or whether any have 'returned' (that is, failed to become pregnant). These markings help Phil's management of the flock as the ewes get nearer to lambing.

So, every day in October someone has to go and catch the rams and freshen the raddle by smearing a bit more between their legs using a flat stick. This is not a difficult job but messy in wet weather when the ram's wool is sodden. It was a task Jethro Larkin used to enjoy. After his death Phil tried to get David to do it but usually ended up having to do it himself. Now it has been taken over willingly by Bert Fry.

David spends most of October surveying the autumnal scene from the seat of a tractor – ploughing, cultivating and drilling. (It's a source of never-ending astonishment to his father that, having been provided with a tractor cab which effectively masks the noise of the engine, David chooses to blast his ear drums most of the day with the radio at full volume. 'Honestly, Jill,' Phil once told his wife, 'he was down re-setting the plough and you could hear the music, if that's what you call it, pouring out of the cab right across the field. He used to complain that the engine deafened him, but that noise must be ten times worse.')

David and his father want to get as much as possible of their cereal acreage sown before they get stopped by the weather. Some years they could go on drilling right through November and even into December. But at the beginning – or even the end – of October they don't know what the weather has in store, so they aim to get as much done this month as they can. Preparing a seed bed for corn is similar to planting in the garden, except that it's more mechanised. First you plough, which is the equivalent of digging, next you break up the furrows with a cultivator, or in the garden a mattock, and finally you harrow, or in the vegetable patch use a rake. An increasing number of gardeners are now using some sort of powered cultivator which does the last two operations in one. On the farm they're taking a similar short cut and David now uses power harrows which in the right conditions can produce a seedbed from the furrows.

David and his father may well spend a couple of hours at the ploughing match arranged by Loxley Barratt and District Young Farmers Club. A local farmer provides a field (and has it ploughed as 'payment') and about 20 local lads plough an acre apiece against the clock. To watch them making their cop (setting up the first furrows), squinting along the top of the tractor then looking anxiously behind at the plough, you would think their very lives depended on the result. In a world full of hijacks, bombs and revolutions, Phil finds it a great palliative to watch these youngsters pitting their skill against one another in burying the golden stubble beneath the rich brown earth.

Perhaps in the second week of the month Phil will decide to lift his 7 acres of fodder beet. This is a root crop, not unlike sugar beet, which they grow at Brookfield for the dairy cows. It has two disadvantages. In wet weather it can be the very devil to harvest, as Phil found out to his cost. A few years ago he was not able to lift 2 or 3 acres because of the mud; luckily he could turn some lambs in on them so they weren't altogether wasted. The other drawback is that the roots need storing over the winter in a clamp which has to be protected with straw and polythene sheeting and needs very careful ventilating. The first year Phil grew them he didn't make a very good job and as a result the clamp heated up and he lost a lot of beet.

To lift the crop he borrows Brian's sugar beet harvester, a fearsome machine, which lifts three rows at a time. This arrangement suits both parties well. Phil doesn't have to invest in a harvester for his small acreage and Brian gets his machine tested under field conditions before he starts lifting his own sugar beet. However well a farmer maintains his machinery there are often time-wasting adjustments and minor breakdowns at the beginning of the season. By lending Phil his harvester along with its regular driver for a couple of days the week before he wants to use it, all the difficulties can be ironed out without holding Brian's chaps up. No cash changes hands; Phil simply drops off half a pig for Brian's freezer as a thank you (without the half-head which Jill included when she tried selling Phil's Middle Whites around the village a couple of years ago).

The final phase of the Brookfield harvest comes a little later in the month when they lift the potatoes. Phil grows about 15 acres which in an average year will produce well over 200 tons – enough to fill more than 150 000 3 lb bags for the supermarket shelves. Although that may seem a lot of potatoes it's not sufficient to justify buying the specialised equipment needed to deal with them. So Phil is a member of a group of local growers who between them own a planter, harvester and a building where the tubers are stored, graded and packed during the winter. Phil has to wait his turn for the harvester but it usually comes around the middle of October, complete with driver and a small team of women whose job it is to ride on the machine and pick out the odd stones and damaged potatoes before they are delivered into the trailer running alongside. A close-knit team, they have developed a ribald sense of humour. The group is always on the verge of 're-equipping' with new machinery which would avoid the need for the gang of women. It can't be too soon for Neil who, often the butt of the teasing, sometimes finds it too much and makes an excuse about a sow farrowing and disappears towards Hollowtree, leaving David or Bert to cart the spuds to the store on a farm about 3 miles away.

Here they are run over a riddle to get rid of the small ones which are unsaleable and not worth the cost of storing. They are then sprinkled with a chemical to stop sprouting and disease and stored in 1 tonne boxes. The humble potato is prey to a legion of diseases, some with highly descriptive names such as common and powdery scab, pink rot, silver scurf, skin spot, gangrene, black scurf and watery wound rot. The same gang of women will be busy throughout the winter grading and packing the tubers, into 25 kg bags for the greengrocers and caterers, others into 3 lb and 7 lb polythene bags for the supermarket.

Three men are leaning on a gate watching the cows in Trefoil, the field across the road from the farm. It's the annual 'committee' constituted to decide when to bring the herd in for the winter – and the cows are very much involved in the decision-making, though it's not quite as

Dairy cows spend about half the year outdoors and the other half inside. When they start treading up the ground like this it is time they came in for the winter.

organised as it seems. Phil strolled over with Graham the cowman to fetch the cows in for the afternoon milking and David, who came by on the tractor, joined them. Phil has noticed that the cows seem to be treading the October pasture up and damaging it. David complains that the last time he milked he spent far too long cleaning the mud off their udders before he could put the teat cups on. Graham remarks that it's not much fun at six o'clock these dark mornings fetching them in from the far end of the field. But it's the cows themselves who finally settle the matter. They just don't look as if they're enjoying life outdoors. 'Look at them,' says Phil. 'Mooching around, searching for somewhere dry to lie out of the wind – staring at the grass rather than eating it. It's time they came in.'

The cows have been out since mid-April, exactly 6 months; living mainly on grass, with a cereals and protein mixture (known as concentrates) fed to them in the milking parlour twice a day. Now they will come into their winter quarters under cover and be fed chiefly on silage. The decision is put off as long as possible for the simple reason that all the time they are outdoors they're finding their own food and spreading their own muck. It's a different matter once they come in.

HOME FARM

Being a mainly arable farmer, Brian finds October a busy month. Not that he takes his coat off very often; his role is more keeping other people's noses to the grindstone. He's the one who's likely to rush into Borchester or further afield for a spare part, if he can't persuade Jennifer to go, that is. He's the one who'll take the extra diesel out to a tractor driver to avoid his having to go back to the buildings and so hold up the job. And he's the one who'll carry on drilling while the driver has a break if they're trying to finish a field before dark.

In addition to his oilseed rape which will already have been drilled, Brian grows more than 1000 acres of wheat and barley, most of which he hopes to plant in the autumn. And since at the beginning of October, like Phil, he has no idea of what the weather will be like later on there is a lot of pressure to keep the drill going. The race to sow most of the cereals at this time of year is of comparatively recent origin. Farmers

Wheat and barley crops sown in the autumn tend to yield more than those sown in spring so most farmers try to drill as many acres as they can before winter sets in.

have always tried to drill as much wheat as they could in the autumn, since a crop which stands the winter usually yields much more than a spring-sown one, but most of the barley was left until the spring. In the early seventies plant breeders produced new varieties of winter barley which yielded up to half a ton an acre more than spring-sown ones and had the added advantage of ripening a month earlier and helping to spread the harvest over the summer months. Their arrival on the scene had a dramatic effect on the countryside. Instead of much of the stubble surviving until Christmas, or even until spring, giving the landscape a more varied appearance and making it easier for walkers and hunting folk to get about, farmers started ploughing every acre as soon as possible after it was harvested and cleared. These days a field of stubble in January usually means that a farmer has slipped up.

To speed up his own autumn work Brian recently invested in a 20-foot-wide pneumatic drill which can sow 80 acres in a good day.

The seed is blown out through a series of pipes so that it trickles into the ground behind the metal coulters which make little furrows for it. At Home Farm they drill 1¼ cwt of barley seed to the acre and 1½ cwt of wheat; they hope to harvest at least 50 cwt of barley per acre and 3 tons of wheat. But there are plenty of potential pitfalls between seedtime and harvest, as Brian and his neighbours know to their cost. Unlike Phil's more traditional drill, Brian's sows only the seed; the fertiliser is broadcast before drilling takes place.

Brian was the first farmer in Ambridge to go in for 'tramlining', a technique developed on the Continent in the seventies. Tracks are left in the crop when drilling to allow other machinery to travel through the crop during the growing season. The 'tramlines' have to be carefully planned so that once established they enable Brian to send a man out at any time to apply a spray to control pests, weeds or diseases, or fertiliser to give the corn a tonic, without damaging the crop itself.

In the seventies new technology threatened the plough, but on most farms ploughing has triumphed. Burning after harvest would have made this long stubble easier to bury.

The fields which Brian drilled in September will be greening up nicely; it takes the seed only 10 to 14 days to germinate at this time of year (longer as it becomes colder). In the old days a farmer would have shut the gate on the crop at this stage and done no more until harvest but, nowadays, when yield expectations are so much higher – and the cash input so much greater – the sensible farmer, which of course includes Brian, will be round his fields regularly throughout the growing season, looking for signs of trouble.

Once the sugar beet harvester comes back from its 'trial run' at Brookfield, Brian can begin to tackle his own 100 acres of beet. It's always difficult to decide when to start lifting since the crop keeps

The longer the sugar beet is growing the heavier the crop and the higher the sugar content. But the weather may deteriorate, so most farmers start lifting in October.

growing right up until the end of the year and the all-important sugar content tends to rise until the beginning of November. If he delayed starting until the middle of November he might well harvest another 2 tons an acre with a higher sugar percentage; on the other hand, no one knows what the weather's going to be like and harvesting beet in the mud is a rotten, costly job. A further factor in deciding when to make a start is pressure from the beet factory to keep up deliveries once they start slicing. But wet weather slows the whole operation down, and when the beet reaches the factory they knock quite a bit off the price for what is termed 'dirt tare' – soil clinging to the roots. Brian still hasn't forgotten the time a few years ago when the weather was really bad and he couldn't get to Smithfield Show at the beginning of December because he was still wrestling with his sugar beet. It was the year after that he bought his new harvester. On a good day it can lift about 8 acres which means that in theory he could deal with the whole crop in a fortnight. It doesn't work out like that, of course; what it means in practice is that now he only lifts in good weather.

Since the crop was introduced to this country on a large scale in the 1920s, sugar beet has become a vital part of the agricultural economy. About half a million acres are grown, mostly in eastern England and the West Midlands, and home-grown beet provides nearly half the nation's sugar. The remainder comes, of course, from cheaper imported cane sugar and the price to the European farmer has to be heavily subsidised by the EEC to make it worth his while. The growing of sugar beet has been revolutionised over the last 30 years by developments in plant breeding, sprays and mechanisation.

'Oh, Brian.' Jenny turns over in bed.

'What is it, darling?' a sleepy voice mumbles. 'Can't you sleep?'

'It's those wretched stags again. Listen.' Nothing happens for half a minute or so and then they hear the long, deep-throated bellow from the deer paddocks beyond the farm buildings. 'There, hear him? It's awful Brian – they keep on and on.'

'Nothing to worry about,' says Brian settling back into his pillow. 'They're only getting on with their job. Now go back to sleep.'

The stags 'roar' all through October; it's the annual 'rut', the period during which the hinds are mated. By the end of the month Brian hopes they will all be in calf. During the day the roaring is not so noticeable as there is usually a tractor running in or out of the yard or the whirr of the grain dryer. The sound can be heard half a mile away at times and several folk have referred to it although no one – except Jenny – has gone so far as to complain. Brian began keeping red deer in 1987 when he sold his herd of beef cows and his Limousin bull. Although he started cautiously with only 25 hinds he is how expanding, attracted not so much by the prospect of producing venison as by the much more lucrative business of selling breeding stock to other farmers.

While the stags are busy securing a crop of calves for next summer, Brian is also organising the mating of his ewes. He manages his large flock of 600 differently from Brookfield. Phil Archer plans to have his flock start lambing at the end of February but Brian splits his into two. The smaller flock of 200 ewes is set to lamb at the beginning of January. This means that they were put to the ram about 12 August, an easy date to remember as it's the opening of the grouse shooting season, the so-called 'Glorious Twelfth'. The rams go in with the larger flock of 400 around 27 October, to start lambing at the end of March.

Both farmers have thought out their lambing strategies very carefully. The price of fat lamb goes up during the winter and spring when there aren't many lambs on the market and comes down with a wallop in the summer when there's a glut. The price is highest from February until May. It would be difficult for Brian to have lambs ready in February since ewes do not naturally come on heat early enough, but he can have lambs born in January ready to sell in April or May before the price falls. It would over-stretch his labour and other resources if all his ewes lambed at the same time, so he lambs the main flock at the end of March. By the time these are ready for slaughter in the summer the price will have plummeted, but all is not lost. When the January lot are born there is no grass growing and they have to be fattened on expensive cereals, but the March-born lambs will have plenty of grass and so will be reared much more cheaply. Phil, on the other hand, takes a middle course. He can get most of the feed for his lambs from grass which begins to grow at the beginning of April and he can sell quite a few lambs before the price hits the bottom in July. Meanwhile both farmers keep the situation under constant review.

BRIDGE FARM

While Brian is driving round in his Range Rover making sure everything is going smoothly, or keeping in touch with his men over a short wave radio, Tony and Pat are bending their backs pulling carrots. At the beginning of the month they are still selling them in bunches with the leaves on but they are getting too big to market like this much longer and soon they will start lifting them mechanically. The tractor pulls a machine through the crop which loosens the carrots and lifts them on to a simple elevator. This shakes off the soil and leaves them lying on the ground ready to be picked up and carted back to be sorted, topped and packed into 28 lb bags. The tops and all the misshapen carrots are fed to the cows and Tony derives a lot of pleasure at seeing them eat the waste which nourishes them and, later through their muck, will enrich the soil.

The size of Tony's potato crop depends almost entirely on how much they have been affected by blight. This is the disease which ravaged Ireland nearly 150 years ago and was responsible for the great famine of 1846. It is still very much with us. 'Conventional' farmers like Phil ward it off by giving their potatoes a protective fungicidal spray but an organic farmer cannot do that. Tony knows he is very unlikely to escape an outbreak and just hopes for the best. If a heavy attack comes early in the growing season it will halve his yield, but a mild attack later on, when the tubers have had a chance to swell, won't be so disastrous. So, it's with some trepidation that Tony digs a root or two in the first week of October and goes back to Pat with the good — or bad — news, although he will already have a pretty good idea of the verdict.

Although he usually takes advantage of the local potato group's planter in the spring, he likes to harvest the crop himself, in between milkings, with the help of a gang of pickers. He uses a tractor-mounted spinner with a share, which passes under the potato ridge to loosen the soil, and a spider wheel which flicks soil and potatoes sideways. The tubers are supposed to hit a net and drop in a neat line while the soil goes through, but its a tricky business adjusting the share to the right height. One moment Pat is stopping him because he's lifting so much soil that the potatoes won't separate out. The next, after Tony has lifted the share, she's screaming, 'Tony, you're slicing them all in half.' In the end they get it right and the pickers put them into buckets which are tipped into bags, loaded on to a trailer and taken back to be stored for a month or two until the price firms. Working a shortened day like this, it takes more than a week to lift his potatoes, especially if it turns wet.

One thing Tony has to keep an eye on during October is weeds in his leeks which cost him half of last year's crop. The difference between leeks and crops such as cabbages is that there isn't enough leaf on them to smother the weeds, and certain species like chickweed just keep on growing through the autumn when most of the annuals have given up. It's a rotten job (the sort Tony would give to Steve, the YTS lad) especially if it's wet and the soil keeps sticking to the hoe.

Later in the month Tony and Pat may cut a few cabbages, making sure they trim them well. A year or two ago they had a load sent back – they had left on too many of the outside leaves which were peppered with caterpillar holes. Tony walks over his acre of Savoys, knife in hand, looking out for any which are fit to cut. He trims them and leaves them for Pat to pack, 12 at a time, into crates. They may cut 20 crates in an hour or two before lunch in this way.

This is really opportunist harvesting as the cabbages are still growing. But they may get a better price for them in October than for a bigger cabbage in November or December. It just depends on the market. By the same token, they may pull some early swedes this month – if the demand is there. The life of a grower – organic or 'conventional' – is different from that of a farmer, as Pat and Tony have discovered the hard way. There are no subsidies or guaranteed prices for most of his produce. He must keep in constant touch with the market and be ready to exploit it. Tony likes wheeling and dealing – he just isn't quite as clever at it as he thinks he is, as Pat points out in her inimitable way from time to time.

He also has to be a farmer, of course. In between harvesting his vegetable crops he has to cope with his cows which need to be milked twice a day, 7 days a week. Like Phil, Tony brings the cows in for the winter about half-way through the month and adds slurry clearing to his daily rota.

November

Although still officially autumn, November can be very wintry – just as October can feel like summer. At times it seems as if winter has taken a firm grasp, with ice on the ponds and everything at a standstill in the fields. Then a few days later you'll see two men digging in a gatepost in their shirt-sleeves with the gnats dancing round them in the thin sunshine.

A hard November is supposed to herald a muggy winter. Walter Gabriel was fond of quoting his grandmother's old rhyme, 'Frost in November to hold a duck – there follows a winter of slush and muck', but like many country sayings it's often stood on its head by the time spring comes.

It's a misty month, though the beguiling mists have a nasty habit of turning into the more sinister fog, as Tom Forrest has cause to remember. He'll never forget the time he lost his way coming back through the woods after an early-morning patrol. It was light and he knew the wood intimately – or thought he did. But he wandered round and round in circles in the thickening mist – or was it fog? He passed the same dead elm three times before finding the edge of the wood and the well-worn path to his cottage, where an anxious Prue had been keeping his breakfast warm for more than half an hour.

The big bonfire in Ambridge on 5 November is echoed in miniature all over the area as gardens are tidied up for the winter and rubbish burned. Meanwhile there's a feast of vegetables waiting to be picked and cooked – cabbage, cauliflower, leeks, carrots, swedes and sprouts – with onions and potatoes galore in the shed. And for those with greenhouses there is also the tail end of the tomatoes and cucumbers, peppers and even aubergines.

It's an unpopular month with some farmers who still have field work to do and aren't certain how long the weather will allow them to do it. St Martin's little summer, which sometimes brings a spell of fine weather around the saint's day on 11 November, may give way to conditions little short of arctic a few days later. If Phil and David knew half-way through the month they were going to have to put their implements away until the spring they could sit back and relax – enjoy some shooting or even go off on holiday. It's not knowing which they find so frustrating.

The livestock farmers on the other hand are settling in for the winter with most of their animals under cover and, they hope, enough feed in hand to keep them until they can go out again.

But if the four-footed denizens of Ambridge are tucked up for the winter the same cannot be said of the two-footed ones. There's an endless programme of winter activities under way: dances, dinners, hunt balls, NFU meetings, Young Farmers' events, whist drives, quizzes and evening classes aplenty at Borchester Tech where the locals can learn anything from how to crochet to the archaeology of the Am Valley.

Cubbing gives way to fox-hunting proper at the beginning of the month, and Shula may scrounge a day off work to follow on one of Chris's horses. The pheasant shooting season gets properly under way with Jack Woolley and Brian Aldridge vying with each other to attract prestigious clients and claim record bags. Meanwhile, folk like Phil will be shooting perhaps once a week on his own or neighbours' land, unlike Brian who may shoot two or three times a week – and much further afield than Ambridge.

BROOKFIELD

When Phil and David bowed to the inevitable and brought the cows in for the winter they knew that they were sentencing themselves to 6 months' hard labour; that until next April they would spend much of their time feeding, bedding and, above all, mucking out their dairy herd. The old saying 'tied to a cow's tail' is usually thought to refer to the twice-daily 7 days-a-week chore of milking, but once the cows are indoors the other tasks are just as regular and equally inescapable.

The 95 Friesians spend their winter in a complex of buildings and yards designed primarily to be as labour-saving as possible. The comfort of the cow also has to be considered because a discontented cow isn't going to milk well, but no one would claim that a cow really enjoys

Cow cubicles give each milker her own 'bedroom' where she can chew the cud or sleep in comfort. At Bridge Farm, however, the herd is 'loose housed' on straw – which makes better manure but risks bullying.

spending 6 months standing on concrete, hoof-deep in its own muck.

At Brookfield the cows are housed under cover in 'cubicles'. These are individual stalls – one for each cow in the herd – bedded with straw, where the animals can go to ruminate or sleep. Their advantage is that the cows can rest comfortably on their own, away from the bullying which takes place in every herd. This system of housing was totally misunderstood a few years ago by an angry correspondent to the *Borchester Echo*. Signing herself 'Disgusted' of Edgeley, she condemned dairy farmers for the inhumanity and greed which led them to 'confine these lovely creatures in barbarous cubicles from which there is no escape'. In fact the cows are free to come and go entirely as they wish 24 hours a day. Most of them adopt a cubicle as their own bedroom, so to speak, and they don't run the risk, as they do in some systems, of getting their teats trodden on by other members of the herd.

It's in the open yards that conditions are not so favourable. The trouble stems from the fact that the cows have got to leave their muck somewhere and the only place they can dung is in the yards or in the passageways between the cubicles. The resultant 'liquid manure' – a mixture of sloppy dung and urine – is known as slurry, a word which has entered the language only in the last 30 years or so since this system became common. Getting rid of the constantly accumulating mass of slurry is one of the perpetual nightmares of the modern dairy farmer. They scrape the yards and passageways every day except Sundays and holidays at Brookfield, but they soon become messy again. To see the cows standing in 2 or 3 inches of slurry on a damp November day is not an attractive sight, although David probably doesn't give it much thought as he has never known it different.

Phil might sometimes cast his mind back to the old days, when he was David's age and the herd spent all winter in the cowshed where they were fed, watered and milked. They only had about a dozen cows at the beginning of the fifties but how Dan, sometimes accompanied by the young Phil, loved to poke his head through the door of the shippon last thing at night, just to make sure that everything was all right before he went to bed. There they would be, his beloved Shorthorns – some red, some red and white and some roan – quietly chewing the cud. Seduced by the warmth, he'd grab a fork and walk along the short row of stalls, stopping now and then to pull a dung pat into the gutter, the sweet aroma – a blend of hay, straw, mangels and muck – in his nostrils. A far cry from today's acrid smell of silage and slurry. Mind you, our friend from Edgeley would have had something to say had she known that they spent the whole winter chained by their necks!

In those days all the feed was carried to the cows by hand and set in front of them, their beds of straw were made up every day and the muck, most of which dropped into the gutter behind them, was forked manually into a wheelbarrow and put on to a dung heap where it gently heated up, rotting the straw in the process. Once a year it would be

The cows steadily eat their way through the silage clamp during winter and become very skilled at feeding above and below the electric wire without getting 'stung'.

spread on the stubbles and ploughed in. How times have changed.

The bulk of the winter ration for Phil's cows is made up of grass silage and they're on what's known as a self-feeding system. They have free access to the silage face day and night, controlled by a single-strand electric fence. If they touch this they suffer a sharp sting and they soon become very clever at feeding under it and over it to leave a remarkably clean face. In addition to silage, once a day they are given a feed of fodder beet which they relish – 'never try to get between a cow and a fodder beet' is the advice Phil gives to anyone involved in feeding them. And twice a day they get a cereal ration in the parlour when they go to be milked.

Basically it's the cowman's responsibility to do all the feeding and cleaning but at this time of year he's too busy. In addition to his normal chores he'll have cows calving, young calves to see to and cows coming on heat and needing the AI man. So one of David's first jobs in the morning, while the cowman's finishing milking and washing-down and the cows are still penned in the dispersal yard, is to scrape slurry from the yards. This he does with great panache from the tractor seat, using a rear-mounted scraper blade. That his mind is not solely on slurry is obvious from the strains of Radio 1 emanating from the tractor cab. He skilfully pulls and pushes and marshals his slurry, eventually sending it up a ramp where it falls into a huge man-made pond known improbably as a 'lagoon'. There it stays until they can get it out on to the land.

Next he takes silage to the young stock – the weaned calves and yearlings – using a hydraulic grab mounted on the fore-end loader of the tractor. This enables him to do two jobs at once. The silage is 7 or 8 feet high in the clamp – and the cows can only reach 5 or 6 feet. If nothing was done they would tend to eat into the clamp and leave a 2-foot overhang of silage which would eventually fall down, perhaps burying the electric fence wire and shorting it. So David 'grabs' the top couple of feet from the silage face to feed the young stock and kills two birds with one stone.

His last job, usually, is to fetch the fodder beet in a bucket mounted on the front of another tractor and tip it into the troughs which run the length of the yard, while the cows are still safely shut away. Whether he stays on after that to help bed the cubicles with straw depends almost entirely on the weather. If it hasn't rained too much and it isn't too frosty – both possible situations even in November – there will always be more field work to be getting on with. Perhaps they haven't yet managed to drill the land from which they harvested the potatoes. So just as David's finishing the fodder beet round, Phil comes into the yard. 'Do you think Lower Parks'll go?' He means will the drill work without clogging up? He doesn't need to spell it out.

'Dunno,' says David. 'What do you think?'

Phil takes his hat off and scratches his head. 'We could give it a try.'

'What's the forecast?' asks David, looking up doubtfully at a sky resembling a wet saucepan lid.

'They speak of rain later on,' Phil replies, 'but we could get it in before then.'

'OK, then,' says David, beginning to take charge. 'You get loaded up with seed and fertiliser. I'm finished here – I'll meet you round there with the drill in half an hour.'

It's late for sowing wheat, but not too late. If they don't manage to get it in this month it will have to be drilled with barley in the spring. Taking average yields this could mean a loss of £100 an acre, or £1400 on that field if they drilled spring barley instead of winter wheat, but Phil and his son wouldn't look at it in those terms. They know that nothing is certain in farming and that the wheat which they were so pleased to get in this November could be flattened in a storm next July, and as a result they could harvest only a third of a crop. Farming teaches its followers to be philosophical in these matters.

The rams are with the ewes at Brookfield until about the middle of the month by which time nearly all the ewes will have been marked with raddle of one colour or another to show that they have been visited by the ram. Phil knows that by leaving the rams in for a fortnight longer, some of those which have not yet been served may still have a chance to become pregnant. But he also knows that a lamb conceived at the end of November won't be born until the last week of April and will be hanging round, still not ready for the butcher, until August when he's in the middle of harvest. Better to cut his losses and have a few barren ewes which will sell well for meat later on. So he takes the rams out after about 6 weeks and puts them in a paddock on their own. The ewes are still living on grass until a November frost comes to put an end to this season's growth, after which they'll need a daily feed of silage. This is dug out with the grab, loaded on to a trailer and forked off in a long line across the field for the ewes to eat. Apart from a daily check they need no special treatment now until the middle of January.

To the pigs up at Hollowtree it doesn't really matter whether it's November or May, the routine's still the same. It's a totally non-seasonal production line, the end-product being pigs tailored for the bacon trade. Although Neil Carter runs the place on a day-to-day basis, Phil reckons to pop in three or four times a week to check that everything's all right. Quite often he gets roped in for a job.

'While you're here, Mr Archer, can you give me a hand to shift those growers?' and Phil finds himself shunting a bunch of 10-week-old pigs along, holding a board in front of him, while Neil goes on ahead to open the gate of the pen and guide them in.

The pig unit was restocked after the calamitous outbreak of swine vesicular disease in 1973, as a result of which all the pigs had to be slaughtered. It consists of 60 hybrid sows, five boars and their offspring of various ages; at any one time there are more than 600 pigs at Hollowtree – quite a responsibility for Neil. The turnover on the unit is more than £80 000 a year, only a small part of which is profit, according to Phil.

A modern pig unit is run almost like a factory; the production line has to be kept going. Neil likes to have four sows farrowing every 10 days. The piglets are weaned after 3 weeks and transferred to special weaner quarters where they spend the next 7 weeks or so. After this they are moved to a building for growers where they stay for about 10 weeks and finally they have about 6 weeks in the fattening house from which they should emerge ready for the bacon factory at about 6 months of age.

As in a factory, any hiccup in the production line can cause chaos. In an industrial process it's normally the shortage of a component; at Hollowtree it's usually what Neil refers to as 'bunching'. 'I got bunching again,' he'll tell Phil Archer. 'I don't know 'ow we're goin' to get round it.' What he means is that instead of having his four sows farrowing at

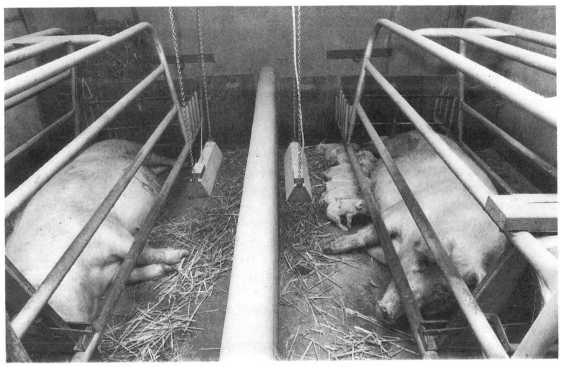

Farrowing crates are designed to prevent sows from lying on their offspring. Saving a piglet often means the difference between profit and loss on a litter. The sow on the left has yet to farrow.

10-day intervals he only had two last time, and this time he'll have six. This throws a strain on the whole system as, buildings being expensive, they have only the minimum number of specialised buildings to accommodate the pigs at the various stages. Neil doesn't know how he'll get round it, but he always does. It usually means putting two or three litters in the old cartshed which isn't really suitable for them and is the very devil to muck out afterwards; or he may have to sell some as porkers at 140 lbs liveweight instead of taking them on to 200 lbs, which is the right weight for bacon. A couple of years ago he got into such a mess that Phil had to sell off a bunch of 10-week-old pigs to clear the system, but that doesn't happen very often.

The trouble is that unlike the factory manager who is dealing with inanimate things like nuts and bolts, Neil's sows have wills of their own and if one decides for some reason or other not to come on heat 3 to 10 days after her pigs are weaned there's not much he can do about it, except complain to Mr Archer.

HOME FARM

With lighter land and better equipment, Brian will be unlucky if he hasn't drilled most of his winter cereals by this time of the year. It's selling the current crop which takes up a lot of his time.

At harvest Brian will be handling something in excess of 3000 tons of wheat and barley plus 200 tons of oilseed rape and perhaps 40 tons of peas. Some of this he stores on the farm and some he sells straight away. Most arable farmers store much of their cereals and spend the winter complaining that they lose money by doing so. Brian's no exception but he knows that he'd lose a lot more if he and his fellow grain producers tried to unload all their corn on to the market at harvest. The reason that he sells some barley 'off the combine', as they say, is that he simply doesn't have enough storage for all of it. So he sells several hundred tons 'forward', by agreeing a price with a merchant in advance of the harvest.

The theory, and normally the practice, is that the price of cereals rises steadily during the winter as the market settles down after harvest. This should more than offset the costs of storage and usually does. The costs include the initial outlay on the grain storage – Brian's set-up would probably take more than £300 000 to replace if it burned down – the running charges and the interest on all the money tied up in stored grain; in Brian's case this is worth over a quarter of a million pounds.

Brian's system, financed out of profits from the years in the late seventies and early eighties when grain growing was a very rewarding occupation, consists of a huge building containing a series of ventilated bins each holding 50 tons, together with an oil-fired dryer capable of handling 20 tons of grain an hour. The whole system is linked by conveyors and catwalks enabling the foreman, to direct corn to any bin or retrieve it for delivery at the press of a button. This copes with nearly 2000 tons, and in addition there is room for 700 or 800 tons stored in bulk in another building with floor ventilation.

It's a highly complex process, drying the corn to the requisite 15 per cent moisture necessary for storage, and shunting each lot to the correct bin. It's not surprising that from time to time things go wrong. Someone has sent a quarter of a ton of valuable malting barley into a bin containing 40 tons of equally valuable seed wheat and an extremely irate Brian confronts the foreman outside the grain store, determined to find out who was responsible. After several attempts to cover up, a very red-faced Steve Manson has to admit that it was the new chap, Ray.

'What the devil was Ray doing fiddling with the controls?' demands Brian.

'I told him to,' returns Steve, 'I can't be everywhere. We had a bullock in the ditch.'

Brian moves out of the way to let a tractor go past before returning to the attack. 'Look, Steve, do you know what that stuff's worth – malting barley and seed wheat?'

'A lot of money, I know,' says Steve. 'Top whack, both of them.'

'Well, it's not worth much now, that's for sure.' Brian's trying hard to keep his temper. 'In fact it'll all have to go to feed the cattle.'

Steve mumbles.

'Now, look here. You go and get Ray – I don't care what he's doing – you go and fetch him and send him up into that bin and tell him to start shovelling. I want all that barley out before I get back and if I find so much as one grain of it left in the wheat, that's his Christmas bonus gone for a burton.' And Brian turns to go.

'He's taken a load of muck into the garden for Mrs Aldridge's roses.' Steve waits to see what effect this has. But Brian whips round.

'I don't give a damn what he's doing – get him up there.' He jerks his head, 'Now.' And adds, 'And tell him to be careful. We don't want to lose more than we need.'

The difference between top and bottom price for a sample of wheat or barley can be as much as £50 a ton depending on its quality. The EEC operates an intervention system for grain. A price is fixed at the beginning of the season, rising each month through the winter, at which grain will be bought in and stored. This goes to form part of the so-called cereal surplus and much is subsequently sold to non-EEC countries at substantial losses. This is a useful safety net for farmers should the price for feed wheat and barley fall too low, but farmers like

Brian would prefer not to rely on it. For one thing the Intervention Board doesn't pay up for 3 months, compared with the local merchant's 28 days. Also the standards have been raised so high in an attempt to deter farmers from looking upon it as an ever-open door for surplus corn that it's far from automatic nowadays to get grain accepted.

The stags have finished their work at Home Farm and stopped roaring, and lost up to 20 per cent of their body weight in the process. The hinds should all be in calf by the beginning of the month. Around the middle of November Brian brings the herd in to its winter quarters – a series of yards converted from cattle pens by splitting them and increasing their sides to 6 feet with wire mesh. Before the deer are put into the pens, where they will stay until next April, the calves must be separated from the hinds and the whole herd treated with a drug to kill internal worms.

The deer contrive to look both timid and slightly menacing as they trot into the yard, led on by curiosity. Brian is constantly improving his handling pens. He has found that the secret lies in moving the deer into progressively smaller enclosures – and the darker they are the better. Brian and Sammy Whipple are both armed with pieces of wood about 18 by 30 inches, resembling police riot shields; these are just in case a hind decides to rear up and 'paddle' at them with her forelegs. They can be quite dangerous and Brian and Sammy are careful not to go in among them at this stage.

The first task is to remove the stags; fortunately they tend to lead the way and are fairly easily diverted into a pen and shut away for the time being. The others are eased through a series of pens until they are packed fairly tightly, then the process of separating the hinds from the calves can be started. The idea is to let the hinds through into the next pen and retain the calves. Brian operates the gate and Sammy Whipple tries to manoeuvre the hinds through – once a few have gone the others will usually follow. Although the hinds weigh about twice as much as their offspring – say 95 kg compared with 45 kg – they're quite easily confused and there's much shouting and swearing as the odd calf slips through or the odd hind doesn't.

'Coming up to the gate, Mr Aldridge,' bellows Sammy, who hasn't really taken to deer. But Brian misses her and then, later, as he lets a calf through by mistake. 'Sammy, why the hell didn't you say it was a calf?'

Eventually they are separated and Brian and Sammy move among them, first the hinds then the calves, one injecting them in the shoulder with a drug to control worms, the other spraying them on the head with a marker to show they've been done. Then they're moved to the pens where they'll spend the winter, the hinds well out of sight of their calves which from now on will have to do without mother's milk and manage on silage and barley meal.

Whenever the weather is favourable the sugar beet harvester is at work. In a normal year they'll finish by the middle of the month, by which time perhaps a third of the crop will have gone off to the factory and the remainder will be piled on a concrete pad for delivery over the next couple of months. By the time they've processed Brian's beet they should have extracted about 250 tons of sugar from it – enough to sweeten one cup of tea for every man, woman and child in the country.

Quite often in early November the 'tramlines' established in the cereals during drilling have their first use. The barley can become infested with aphids which spread a devastating disease called barley yellow dwarf virus. Aphids – greenfly or blackfly to the gardeners – feed by puncturing the leaf and sucking the sap. If the plant is infected it's easy for them to spread the virus. If they do appear it's not long before one of Brian's chaps is threading his way up and down the tramlines with a spray boom covering 24 metres at a time.

BRIDGE FARM

'Pat!' Tony's anguished yell reverberates across the yard. It's coming from the old cowshed, now used for calf-rearing.

'P-a-t!' The cry is repeated, more drawn-out this time. Pat, who is mucking out the stable, listens, screws up her face for an instant, and carries on shovelling. She has heard Tony's shouts before; usually he simply wants her to fetch something.

'Pat! For Pete's sake,' comes the agonising call a third time.

'Coming.' Pat decides she ought to go, leans her shovel against the door and marches across the yard. For once it isn't a false alarm. Tony is in a loose box holding a frantic heifer at bay while her new-born calf lies sprawled against the wall.

'Quick. Get the calf out,' screams Tony as the heifer makes another lunge at it. Pat needs no second bidding and in a few moments has dragged the inert little body into the alleyway. Tony follows her out, bolting the door top and bottom. They both stoop to inspect the battered bovine baby and to their joy find it not only miraculously alive but apparently unharmed.

A few minutes earlier Tony had popped his head over the door of the box to see how the heifer was faring. He knew she was near to calving. What he saw, to his horror, was the heifer savaging her newly-born calf, scooping it up with her head and flinging it against the wall of the box. He tried to rescue the calf but wasn't able to on his own. All he could do was to grab a stick, beat the heifer off – and shout for his wife. Talking about it later he remembered the time a first calver had put Dan *and* her calf out into the passage at Brookfield. Such behaviour is thought to be hereditary and Tony makes a mental note not to breed from the heifer again. When she has calmed down he'll put her back with the herd to go through the milking parlour with the others. Fortunately another much older cow has calved earlier in the day and after a little persuasion she allows the young calf to suckle along with her own, so all is well.

November is one of the peak months for calf-rearing at Bridge Farm. Most of Tony's 65 cows calve between September and Christmas. The 'maternity unit' is in the old cowshed, where the cows were housed and milking done in the days before the parlour was built. It has been converted into four calving boxes, about 25 individual pens for calves and a 'sick bay' with a heat lamp for any calf which is off-colour. When a cow looks near to calving she is put into one of the boxes and left to get on with it. In a normal birth the calf is presented with the front feet appearing first and the head following, lying neatly between them; in most cases like this the cow will calve naturally on her own. Sometimes, with an exceptionally big calf, she may need help. Occasionally a calf is lying the wrong way – there may be only one foot coming or no sign of the head – in which case assistance from the farmer or the vet will be needed. That's why, once a cow is in a calving box, Tony or Pat are never too far away. A dead calf could lose them £200 and a dead cow £600 or so. They know they mustn't interfere too soon, but they'd be equally foolish to leave it too late.

Once the calf is born it's left with its mother for 4 or 5 days. This is to make sure that it gets plenty of colostrum, the first milk, rich in protein and containing antibodies which help to ward off disease. Because the cow is giving far more milk than the calf needs, she is usually taken away on the second day and put through the parlour with the others, but brought back following the afternoon milking and left with her calf all night. On the fifth day the calf is put into an individual pen and the cow returned to the herd. For the next 2 months it's fed

Scraping slurry, the never-ending winter chore on intensive dairy farms. The sloppy mixture of dung and urine is usually pushed into a 'lagoon' until it can be spread on the land. At Bridge Farm, however, they compost their slurry with straw.

twice a day on whole milk from a bucket, but at the same time it's encouraged to eat hay and a cereal mixture. Tony and Pat have changed their system of calf-rearing since they became fully organic. In the past, once the calf had had its initial suckling period, it was fed on a milk substitute and weaned off this on to solid food at the end of a month – the system still practised at Brookfield and in most other commercial dairy herds.

At about 9 weeks – only 4 or 5 in Phil's case – the calves are weaned and transferred in groups to open-fronted sheds, where they live on hay and silage plus a cereal mixture until they are old enough to be turned out. In the case of calves born in the autumn this will not be until the following May.

Normally Pat looks after the calves every morning while Tony hoses

Vegetable growing brings the farmer much nearer the sharp end of the market than producing crops like cereals or sugar beet. There may be more profit in a half-grown cabbage in October than a perfect one in December.

the parlour, scrapes the slurry and feeds the young stock. This means that in the absence of other work they can both start harvesting their crops at this time of the year by about 10 o'clock. Depending on the job in hand they may have recruited some casual workers, although they try as far as possible to spread the work out so that they can do most of it with their own labour. For instance a larger grower would tend to harvest all his cabbages at one time, but Tony and Pat go through them every week at this time of year cutting the best ones for the co-op or for sale locally.

The co-op lorry collects on Tuesdays and Thursdays and most of their work in the fields is carried out on Sunday and Monday, getting produce ready for Tuesday; and Tuesday and Wednesday preparing a load for the Thursday collection. Fridays and Saturdays are slacker although they can always get on with grading and packing the carrots which, unlike the cabbages, do not have to be marketed fresh. They are never short of a job.

In November they will start lifting the first of the leeks. Tony drives the tractor through them drawing an undercutter, a sort of blade which passes underneath to loosen them. He and Pat then pull them by hand, knock them on their gumboots to remove as much soil as possible, trim the excess leaf and cart them back to the buildings where they are trimmed again, hosed to clean them up and packed into 12 lb boxes. A normal week's output for the co-op during November would be:

30 boxes of leeks @ £2.80 a box	£ 84.00
50 boxes of Savoys @ £2.50	125.00
35 nets of Dutch White cabbage @ £3.50	122.50
80 bags of carrots @ £3.75	300.00
Total	£631.50

In addition to this, if there's a strong market, Tony may be tempted to lift a few swedes but he'll be in no hurry as they stand the frost quite well. He'll be keener to get the carrots into the building, remembering the time when he left it too long and lost a third of his crop in a December frost.

At the beginning of the month Pat's free-range hens will still be gleaning on the wheat stubble but laying very few eggs. This is partly because they are only just beginning to get over the annual moult, when they lose and then re-grow some of their feathers, and partly because the days are getting shorter. So, early in the month Tony will tow their house, which is on wheels, into the yard and connect it up to a large run fenced off in a corner of the orchard.

It would probably have been more cost-effective to have brought them in earlier but Pat likes to see them fanning out over the stubbles, finding much of their own food from corn left by the combine, for as long as possible. However, once the hour has changed at the end of October, walking over each evening in the twilight to shut their pop-hole becomes less attractive, so in they come.

The other reason for bringing them within reach of the buildings is that, naturally, hens tend to lay only during spring and summer, and if they were left in the fields all winter egg production would dwindle to nothing. The laying process is triggered off by a lengthening of daylight and it's possible to produce this effect with artificial light. So, once they are back at the farm, Tony fixes up a light on a time switch which comes on in the night and deceives the birds into thinking that spring is on its way. Before long they are laying again.

December

The calendar year moves inexorably towards its close, gathering with it as the month progresses such seasonal happenings as Royal Smithfield Show, the holly and mistletoe sales, Borchester Fatstock Show, the shortest day, Christmas, Boxing Day and New Year's Eve. Although snow scenes proliferate on the greeting cards, Christmas is far more often green than white and the month as a whole can be quite open. Heavy snowfalls are not common, though frost can be a frequent visitor.

On most of the farms in Ambridge it's a question of getting through the winter routine and keeping up with the social round. There isn't time for much else, as the light goes soon after 4 o'clock in the afternoon. With the worst of the weather still to come, it's a restful month with plenty to do but no great sense of urgency.

One circumstance in which Brian and Phil may find themselves spurred to sudden action is an incursion of wood pigeons on to their oilseed rape. Huge flocks of these blue-grey birds – attractive enough on their own but quite menacing in their hundreds – migrate here at this time of year and can eat a promising crop of rape right down to the ground. It's not long before the bird scarers are set up. These are usually bangers operated by bottled gas which explode at pre-set intervals. Each time they go off the birds lift into the air – and then settle again. They're still laughing down at The Bull about the Snells. Robert and Lynda came in one night with the story that the birds had become so used to the bangs that they could actually anticipate them. Yes, they said, they had watched them from Lakey Hill. They definitely took off *before* the explosion, each time. Tom had to explain gently, as he'd had to do on previous occasions, that it took the sound a second or two to travel from Phil's rape field to Lakey Hill, whereas the pigeons had heard it straight away.

One December night six or seven years ago a white-faced Elizabeth staggered downstairs at Brookfield, convinced that someone was being murdered. She'd been watching a late-night film in which a badly injured man had been desperately clawing his way towards a house and succour. Imagine her panic when she reached her bedroom and thought she heard him in the garden, his strangled screams getting closer and closer. An extremely sceptical Phil followed her back upstairs to her darkened room. After a few seconds it came again – an unearthly shriek from the orchard. For a moment or two Phil stood transfixed. Then he relaxed and smiled reassuringly. 'You know what that is,' he said. 'It's a vixen. Listen.' It wasn't long before they heard it – the short, sharp bark of the dog fox which was courting her. As if to complete the night's eerie symphony, a tawny owl hooted right outside the open window.

BROOKFIELD

Phil bustles through the kitchen, obviously on his way out. He's wearing a suit so the odds are that he's heading for the magistrate's court.

'Seen my car keys?'

'Aren't they on the hook?' asks Jill, up to her elbows in flour.

'No – don't seem to be,' he replies. 'Oh, here they are in my pocket all the time. Look, there's two cows for the AI chap when he comes. Graham's away. Both Friesian bull of the day. OK?' And he's off to administer justice in Borchester.

Phil's already given all the information to the local AI Centre's answering machine; he doesn't trust it entirely so he's just making doubly sure. Getting his cows back in calf on time is just about the most important job facing the dairy farmer. Feeding them, milking them, even calving them, are easy by comparison. If he misses a cow's heat period he has to wait 3 weeks for the next opportunity and the experts have worked out that it loses him about £60 a time. With nearly 100 cows in his herd he mustn't slip up too often.

Apart from the cost of every missed opportunity, with an autumn-calving herd and a gestation period of 9 months he needs to get each cow into calf within 3 months of her calving so that she calves again within 12 months – otherwise he soon ceases to have an autumn-calving herd. So, if a cow calves in September this year and isn't inseminated until January she won't calve until October next year; and if she then isn't put in calf until February she doesn't calve until November the following year. It's not long before it's a spring-calving herd. Graham Collard, the cowman, receives a bonus of £2 for every cow with a calving index of 365 days or less to encourage him to spot each one as she comes on heat, or 'bulling' as he'd call it.

When a cow's bulling she looks unsettled; she allows other cows to mount her and sometimes tries to mount her fellows. At this time of year Phil goes round with a notebook in his pocket, writing down the numbers of any he suspects. 'I think 221's coming on,' he'll say to Graham, who promises to keep an eye on her. There could be half a dozen by the end of the day – two he's certain of, two he's fairly sure are on heat and a couple which could be. Dan Archer's last job each night when he was Phil's age was to look into the shippon to make sure his precious Shorthorns were all lying down comfortably chewing the cud. Phil's chore, before he turns in at this time of year, is to slop through the slurry-covered yard to see if he can spot a cow 'coming on'. Down will go her number in his little book and, if a further check in the morning confirms his suspicions, she is diverted into a pen after milking to await the attention of the AI man.

At Brookfield they keep a 'commercial' herd of Friesians as opposed to a pedigree one; they're interested in selling milk rather than selling breeding stock. That's why Phil has the 'bull of the day'. He's not especially interested in its name or breeding, but he knows that the AI Centre will have made sure that it's a good one. He puts about 60 of his 95 cows to a Friesian bull; this will guarantee him the 25 heifers he needs for replacements. The bull calves will be sent to market when they are a week or 10 days old and will be reared by someone else eventually to produce beef. The other 35 cows in the herd will be mated to beef bulls. Phil can choose from a list of foreign breeds now established in this country – Charolais, Limousin, Simmental, Blond d'Aquitaine, Belgian Blue – as well as native breeds like Hereford and Aberdeen Angus. Obviously he is careful to put his better cows to the Friesian bull, to breed his herd replacements, while the poorer yielders or ones with undesirable characteristics are mated to beef bulls so that their weaknesses aren't perpetuated in the herd. They do keep a real bull at Brookfield, a Hereford which is run with the heifers to produce the excellent beef-cross which is the familiar black animal with a white face.

On most dairy farms calves suckle their mothers for a few days after which they are given a cheaper milk substitute before being weaned on to dry feed at 4 or 5 weeks.

Sometime during December Phil is almost certain to receive a card from the secretary of the South Borsetshire Hunt informing him that hounds will be meeting in the vicinity. If he has any objection to their coming over his land he is asked to telephone or write. This is a clever combination of inertia selling and moral blackmail which works effectively as comparatively few farmers in the Ambridge area do deny access. Phil's attitude is typical: he doesn't hunt himself but his daughter does; he quite likes seeing the hunt – it's part of rural life; he wants to see the foxes controlled but not exterminated, and neither horses nor hounds have done him any real harm in recent years. It's true there was that time they broke those rails between Long Field and Wormitts but the following day they sent some members of the Hunt Supporters' Club to replace them. And he can remember when a rider, too idle to get off his horse, left a gate open and the sheep got into the barley in Ashfield; but several of the foot followers turned them out and the Master rang the same evening to apologise. He can't deny that they tread the corn up a bit round the headlands but they probably don't do anything like as much damage as the badgers do later on, rolling around in it, or the rooks when the barley's in ear.

The moral argument, pursued so vehemently elsewhere, is rarely articulated in the Brookfield kitchen. They tend to accept hunting as a fact of life, like November mists and the 'Min of Ag'. The biggest turnout of the year is at the Boxing Day meet, usually held in the square in Borchester. Helpers from The Feathers take a glass of hot punch to everyone who's mounted and quite a few who are not. It attracts not only hunting folk but scores of townspeople and visitors and is regarded by the hunt, keen for popular support these days, as an opportunity for good public relations. No one expects an exciting day's hunting but the Master tries hard to put on an attractive spectacle for the hordes of car followers. The sight of horses and hounds, red coats and the occasional top hat will set the cameras clicking, and if they catch a glimpse of Charlie slipping down the side of a hedge and hear the horn as the huntsman lays his hounds on, their day will have been made. The regulars won't mind too much if the once-a-year followers head the fox; they know that it's all in a good cause and next week things will be back to normal again.

Meanwhile, Brookfield has settled firmly into its winter routine. Milking, feeding, mucking out, the daily round at the pig unit, taking silage to the ewes, sending off the odd load of corn – and coping with the unexpected. As Christmas approaches Phil and David turn their minds to making life as easy as possible over the holiday. Bales of hay and straw are left at strategic points round the yards so that all that's needed on Christmas morning is to cut the strings and throw them over the rails. The yards are scraped specially clean, the silage face made extra tidy. Graham Collard and Bert Fry have both had an extra £20 in their wage packets and when they knock off on Christmas Eve they're invited into the Brookfield kitchen for a drink. Jill passes round the mince pies and Phil charges the glasses as both sides try to feel at ease. Outdoors their relationship is perfectly natural but in the house, even though it's Christmas, the conversation is slightly forced as they exchange pleasantries and the odd leg-pull.

When they have gone Phil falls to thinking of Christmas at Brookfield when he was a lad. It wasn't £20 in the pay packet in those days – farmers like Dan Archer had little money to spare in the thirties and the farm worker earned less than £2 a week. Their Christmas boxes were of a more practical nature – for those with young children Doris Archer would make clothes with remnants bought in the market and for each man there was half a pig. Some of the men had their own styes but Dan didn't encourage them to feed a pig; he suspected that they would often be kept on his barley meal, and while he didn't grudge the meal he felt it was bad for discipline to condone stealing.

Phil remembered helping on Christmas morning when all the staff came in for a few hours – to see to their cart-horses, to milk the cows by hand, to feed the pigs and take hay to the ewes. Farming was all so labour-intensive in those days. Then came the turkey and Christmas pudding and in the evening neighbouring farmers like Jess Allard and Fred Barratt would come over with their families, in horses and traps, and everyone had an evening of innocent pleasure; the children larking about, the men playing halfpenny nap and the womenfolk gossiping and giggling as they prepared the supper. The three families got together regularly during the winter to play cards and normally drank nothing stronger than home-made cider. But Dan would buy a bottle of cheap sherry or port-style wine at Christmas and there were plenty of jokes about 'Steady on there – you'll have us all in the ditch' or 'Good job the horse knows his way home' as they were offered a second glass. And on Boxing Day they all went rabbiting with ferrets, the children gathering up the rabbits as the men shot them. And in the evening, more cards – this time at one of the other farms, usually Jess Allard's.

On livestock farms, these days, the work on Christmas Day and Boxing Day tends to be carried out by the family, giving the men a break and saving on overtime. Arable farms tend to close down for about 10 days, although at Home Farm Sammy Whipple has to come in and see to the stock. Brian knows that by the time they've taken out Christmas Day and Boxing Day plus a weekend plus New Year's Day he isn't going to get much work done in between. Far better to encourage his chaps to take some of their 4 weeks holiday entitlement for the intervening days and let them enjoy a good break. At Brookfield, with more stock, it's not so simple but they manage to feel quite virtuous on Christmas morning as David does the milking while Phil feeds the pigs at Hollowtree. They co-operate over the rest of the feeding – sheep, calves, heifers – before going in for a late breakfast. Phil usually copes with the afternoon milking, often carried out in a mild alcoholic haze and timed to coincide with that film on television which Jill is determined to see but which he is equally anxious to avoid.

On Boxing Day it's more milking and feeding, with a day's shooting at Home Farm sandwiched in between and a party at Brian's in the evening to round it off.

HOME FARM

Most of the ewes at Home Farm will be under cover now. Brian erected a huge building for them about ten years ago. Housing sheep during the winter is a comparatively new concept since they are well adapted to living outside and, except in very severe weather, are better off in the fields. The two main advantages in bringing them in are that the sheep don't tread the land all winter (so the grass comes earlier in the spring) and that it makes shepherding easier. They live in pens of 40 or 50 either side of a central passageway wide enough to take a tractor and trailer which feeds silage into troughs. As they near lambing they receive a ration of concentrates each day. Two hundred of them are due to start lambing in the first week of the New Year so, as the month progresses, Sammy Whipple keeps an increasingly close eye on them.

The shooting season reaches its peak in December and Brian holds a shoot once a fortnight. Since he had his row with Jack Woolley over the employment of George Barford, which ended in the sacking and subsequent re-employment of George and the end of the ill-fated 'shooting weekends', Brian has gone in for shooting in a big way at Home Farm. The season is a comparatively short one. Although it starts officially on 1 October it doesn't usually get properly under way until well into November; it tails off again in January and finishes on 1 February. The reason for this is purely the management of the pheasants. In October the young birds are not big enough to provide good sport and in January they don't want to deplete the breeding stock too much or it will mean fewer birds the following season. Running a shoot Brian's way is big business – the costs are high but the rewards are very attractive.

Left to their own devices, wild pheasants wouldn't provide anything

'Tramlines' enable fertilisers and pesticides to be applied throughout the growing season without damaging the corn. They are created at sowing time by leaving undrilled strips.

like enough sport for the sort of guns Brian attracts, even on a farm like his of nearly 1600 acres. So, early in the year they trap a number of hen pheasants together with some cock birds, pen them up and keep on collecting the eggs – they'll lay as many as 40 per head – and hatching them in an incubator. It's a very skilled and time-consuming job but it means a huge increase in the number of young pheasants, especially as

The processing factories cannot take all the sugar beet as it is lifted, so it is stored on the farms and sent off later. A cleaner-loader helps to remove unwanted dirt.

the hens, when released, usually go off and hatch another clutch of eggs in the hedgerow. Sometimes even this fails to provide enough birds; that's when a shoot buys in day-old or older birds from a game farm. George Barford doesn't approve of this; he calls it 'chicken farming'. Brian is not so fussy. He needs 3000 pheasants and if his keeper cannot hatch them all he's not averse to buying them in.

Once the birds are reared the keeper's task is far from over. He has to make sure that they stay on the shoot by regular feeding – Brian's eat about 20 tons of corn a year – and by creating an attractive environment for them. Home Farm is already well-supplied with woods although Brian is planning to plant more – taking advantage of government incentives for farm woodlands. He also plants strips of maize and other crops which pheasants like along the edges of some of his fields or in awkward corners.

The pay-off for all this effort and expense comes when the shooting season starts. Some companies like to treat their clients to a day's

Shooting is becoming big business on some farms with guns prepared to pay £300 for a day's sport. On other holdings it continues to be a congenial way of spending a few winter days with friends.

shooting and will pay £15 per bird shot for the privilege; if the day's bag is 250 pheasants, Brian collects £3750. And the keeper also does very well in tips. On other occasions 10 guns will each pay £300 for a day's sport. Sometimes Brian will shoot with his guests but he usually prefers to 'organise' the day, staying in touch with his keeper on the walkie-talkie, making sure they keep to time and seeing that the lunch is ready. If everything goes well – and with Brian in charge it usually does – his guests depart at dusk carrying a brace of pheasants apiece, well pleased with themselves.

Boxing Day is reserved for friends, most of whom will have shoots of their own and will already have given Brian a day earlier on. One or two useful people who can't reciprocate in kind may be invited – like his bank manager – and later on, in January, there'll be a shoot for the staff. This is partly because, although he shouts at them occasionally, he has a very high regard for his men and genuinely wants to give them a good day's sport, and also because he wants to repay them for the help they have given to the shoot during the year. Perhaps they'd noticed a pheasant's nest or a suspicious character around the farm. But by this time they'll be instructed to shoot only cock pheasants. As the end of the season approaches, the cocks need thinning out (one cock to five hens is a good ratio) and they want to preserve as many hens as possible to breed next season's birds. So there'll be shouts all day of 'cock over' – and woe betide anyone who bags a hen.

BRIDGE FARM

The lorry from the organic co-operative appears to have a voracious appetite. Twice a week it rumbles into the yard and carts away everything which Pat and Tony have managed to get ready: leeks, carrots, Savoys, Dutch white cabbage or swedes. Sometimes the co-op will ring up and ask them to send more of one vegetable or less of another; occasionally they'll telephone to complain of the quality. But usually the demand is good and the amount of produce ready for collection when the lorry arrives is limited only by the time available for harvesting and packing. The potatoes are stored in the barn away from the frost and they sell these either in bulk, several tons at a time for pre-packing by the co-op, or in 25 kg bags after they have been riddled on the farm. Also stored in the barn are the carrots — these are topped and sorted and packed in 28 lb bags. The cabbages, leeks and swedes are still in the field although some time this month Tony lifts the swedes and brings them into the buildings together with the Dutch white cabbage. Tony and Pat like to send away as much as possible ready for the market as they receive a higher price for it. The co-op deducts 15 per cent of the price realised for cleaning and packing.

The 15 acres of wheat on Bridge Farm has yielded over 25 tons of grain suitable for breadmaking. Most of this is sold to wholesalers specialising in organic flour-milling; it makes 50 per cent more than ordinary milling wheat. Tony takes some of it, about half a ton at a time, to a friend with a small mill who grinds it and packs it into 25 kg bags for local health food shops. Some of this Pat re-packs into 3 lb bags for sale at the back door and to friends. She charges 75p a bag for this and Tony sometimes contemplates wistfully how marvellous it would be if they could market it all in this way; it would bring in £560 a ton, more than three times the price they get for selling the wheat in bulk. However, as she weighs it into bags on the kitchen scales, while the children are clamouring for their tea, Pat is not quite so sure it's worth the effort.

Borchester Christmas Fatstock Show is traditionally held in the second week of December, and in recent years Tony has become interested in showing a beast or two in the hope of winning a prize. No one can remember the date of the first show of its kind but records go back to the middle of last century. There's an engraving hanging in The Feathers of the well-supported show of 1868, depicting rows of beef cattle tethered in the square with sheep and pigs penned alongside. Now it's held in the cattle market and organised by its auctioneers. In these days of wholesale butchers and supermarket meat there are those who would write it off as an anachronism, but a visit to Borchester market on the day of the show would demonstrate the interest which still prevails.

As long as there are farmers producing beef cattle, lambs and porkers there will be those who want to prove that they can turn out a better animal than their neighbour. Unless he wins the championship, Tony knows that there's not a lot of money to be made. But there's a lot of pride in bringing home a prize card and, of course, if he doesn't manage to win it won't be his fault; he'll tell Pat it was because the judges didn't like that particular cross-breed.

Tony finishes milking early on the morning of the show and soon after 9 o'clock he and Pat are in Borchester market. Although she teases him about his show beasts she's really as keen as he is – after all she did her own share of showing before she met Tony when she was running a herd of Welsh Black cattle in mid-Wales. By this time they have decided which of the cattle they've been bringing on will be entered; there's a wide choice of classes – for single animals, bullocks or heifers, for pairs and for five bullocks or heifers. There's also a class for smaller farmers (under 100 acres) and for exhibitors under 21. Tony and Pat usually end up showing a single bullock; a pair need to be very evenly matched to stand a chance. Soon they're busy shampooing the beast, blow-drying it (with Pat's cordless hair drier), curry combing it and fluffing out its tail before the judges get to work at 10.30 am.

There is never a shortage of winter jobs on a farm. Gates need to be re-hung, fences mended and culverts unblocked. In reasonable weather it can be very satisfying work.

Then it's nail-biting time as the judges — usually a butcher and a farmer — move from pen to pen conferring in undertones. Eventually all the cattle are judged, and first, second and third prizewinners selected in each class. Then follows the parade of winners as the judges select a champion and reserve champion. These are the real victors of the show — the ones which will send their exhibitors home with a handsome silver cup and a fat cheque. The champion exhibitors tend to be a self-perpetuating elite of which Tony is unlikely ever to become a member, but this doesn't worry him. A prize card of any sort would delight him — to be nailed up in the dairy — and there is still considerable pride for him in honourable defeat.

After the judges have finished and everyone has had a chance to scrutinise and comment on their decisions, the entries are sold by auction. This is where the prizewinners reap their reward. The prizes themselves are modest; £10, £5 and £3 for the first three in each class, hardly enough to pay for the petrol to bring them to the show. But a first prizewinner might be expected to make 10 or 20 per cent over the commercial price — perhaps £575 for a bullock which would normally have sold for £500. And the champion — always sold at a pre-arranged time so that no one will miss the highpoint of the day — has been known to make £2000. The butchers who pay these prices exhibit the prize cards in their windows in the pre-Christmas period and anyone buying a piece of their beef can always imagine it came from the winner.

If Tony has won, the prize money disappears immediately after the sale in the market bar. If he hasn't won, he joins his friends in the bar to drown their sorrows and question the judges' skill.

The following week usually finds Tony in another cattle market — though this time the pens aren't filled with cattle but with holly, mistletoe and Christmas trees. Tenbury, in neighbouring Worcestershire, is a mecca for buyers and sellers of this seasonal produce and prices in recent years have been high enough to tempt even the most reluctant farmers into searching their hedgerows and orchards. Holly has made £50 per cwt and mistletoe as much as twice that. Tony has found that the proceeds of his annual trip to Tenbury with a horse-box full of holly and mistletoe go a long way towards buying his Christmas presents and accompanying pleasures.

January

This is the month when the countryside seems at its most dead, especially when there's a hard frost to emphasise the feeling. Nature slumbers and the very grass in the fields is brown and withered. Yet, later in the month, the snowdrops in the sheltered garden of Glebe Cottage will break through the ground to prove once again that it was a brief sleep. It's trite though true to say how impossible it is to generalise about the weather in this country. Ambridge has languished under 2 feet of snow at the end of January some years and yet, not so long ago, the daffodils were in bloom at Grey Gables before the month was out. Even in a 'normal' year, whatever that is, some shrubs like honeysuckle will break into leaf and it's not unknown for the rising lark to sing in the wintry sun on Heydon Berrow.

The feeling of another spring in the offing – albeit a couple of months away – visits Tom Forrest as he looks through his seed catalogue and decides what to order. Early potatoes of course, to get them well-chitted before planting, and perhaps he'll try some of that celeriac; Mr Fletcher from Glebelands gave him one of his roots last backend and it did wonders with Prue's stews. Everything looks so enticing in the catalogue round the fireside in January. Oh yes, he'll send for some hybrid lilies for Prue as a surprise – she'll like those. And if there's a chance he'll dib in a row of broad beans before the end of the month.

To make up for a dull and, at times, depressing season the social calendar is even fuller than usual. True, the weather can interfere but it's amazing the lengths to which country folk will go to attend a favourite party or dance. A four-wheel-drive tractor took half a dozen young farmers, seated on straw bales in the transport box, to a ball a year or two ago when the roads were snowed up. Highlight of the month is Borchester Agricultural Society's annual dinner where guests eat the best beef of the year. It's selected on the hoof at the fatstock show and hung for 5 or 6 weeks to improve flavour and tenderness; each table of ten has its own joint carved by a member of the committee.

Shooting tails off during January with a few 'cocks only' days for tenants, neighbours and staff. Hunting continues throughout the month although there are often days when the followers hunt on foot because it's too frosty to ride or because it's unduly wet and would cause damage to the land. These days often end up in a good party back at the pub or private house where they met, as followers don't have to box their horses home and tend them.

It's not unusual to see huge flocks of starlings at dusk this time of year, wheeling and diving in their thousands as they home in on their chosen roosting sites. Woodland owners pray that they will steer clear of their young plantations as the very weight of the birds and their droppings can wreak havoc. They're still talking in The Bull of the attempts to shift them from some woods between Ambridge and Loxley Barratt thirty years ago when pest officers tried firing rockets into their roosts as they settled. The resultant 'explosions', as tens of thousands of starlings soared into the air, were extremely spectacular and attracted sightseers from miles around. The long-term effect was, however, less dramatic. Most of the starlings settled back in the woods afterwards and the others went and spoiled someone else's plantation.

On the livestock farms the daily routine keeps everyone busy. On the arable farms it's sometimes difficult to find enough for everyone to do and Brian is always glad if any of his chaps want to take part of their annual holidays this month. Indeed, Brian himself has been known to take part of *his* annual holiday once the worst of his early lambing is over – and make for the ski slopes.

An explosion of starlings is quite dramatic but usually bad news for farmers. The birds savage corn and fruit – especially cherries – and in large numbers can kill off whole plantations of young trees.

BROOKFIELD

If there is going to be snow in Ambridge it usually arrives this month or next and to a farm like Brookfield it can bring utter chaos. Arable farmers like Brian Aldridge don't mind so much. In fact they quite welcome a covering of snow on their crops; it protects them from the frost and wind. It's quite different on a holding like Phil's with pregnant ewes in the fields, a bunch of heifers lying out at the other end of the farm, the pig unit separated from the main steading and a dairy herd producing around 2000 litres of milk a day. Especially if it drifts or freezes.

'It's trying to snow,' says Phil as he comes in about 5 o'clock, striving to sound light-hearted about it, though secretly full of apprehension.

'Yes, they said it might on the wireless.' Jill's busy at the Aga. 'Like a cup of tea?' They look at each other for a second, each knowing what the other is thinking.

Later they watch the television news. Already a blizzard further north has brought trouble on the roads and more snow is forecast, but the weatherman is uncertain as to how much. There's not a great deal Phil can do about it except hope that it won't be too bad in the morning. He telephones his cowman more to exchange reassuring words than to issue any orders – remarks like 'maybe it won't be too bad' and 'keep our fingers crossed' seem to dominate the conversation; then he rings Neil. Neil's already up at Willow Farm taking some sort of evasive action with his own outdoor pigs; Susan hopes he'll be able to get to Hollowtree in the morning but she can't promise. 'Keep our fingers crossed, then. Good night,' says Phil as he puts the telephone down. He goes out to have a look at the cattle and is relieved to find that it's only snowing lightly. Maybe it won't get any worse.

But it does. When Phil peers out of the window across the darkened yard at 6 o'clock the following morning, he can hardly see the outline of the parlour. The snow has swirled round the yard and drifted half-way up the walls. It's still snowing but he's relieved to see a light on in the parlour. Graham must have got in through the collecting yard; he certainly couldn't have got through the door from the yard. He dresses quickly and hurries across, carrying two cups of tea – one for Graham and one for himself. As he trudges across the yard through a foot of snow he's comforted by the sound of the vacuum pump. At least everything seems to be working there. Graham already has the first cows in – the rest of the herd is standing motionless outside, deep in a mixture of snow and slurry. As he chats to Graham he hears the sound of a tractor; it's David on the four-wheel-drive monster.

'It's hopeless,' he reports. 'It's drifted 6 or 8 foot high in the lane. I've been trying up the other way but it's just as bad. The milk lorry'll never get through.'

'We'll have to use the emergency tank,' says Phil. 'I'll ring Harold in a minute and see what's happening.' Most dairy farmers possess a tank which can be loaded on a trailer to take the milk to a pick-up point on a main road. To avoid having every farmer ringing the dairy about milk collection, there's a contact man in each area; in Phil's case it's Harold Baker who farms on the Hollerton road. He confirms over the telephone that the tanker will collect from Wharton's Garage at midday or as soon after as the driver can make it.

There are two major worries for a farmer like Phil during snow. One is how to get his milk away – the cows carry on producing as usual and if the tanker cannot get to the farm, the farmer must get to the tanker. In Phil's case there's well over £300 worth of milk a day to be got to the dairy. The other concern is whether he has enough feed for his livestock if he's unable to obtain fresh supplies. The cows certainly won't starve as they have plenty of silage on hand, although their yields will drop if they run out of dairy nuts. Sheep are amazingly tolerant and will pick up enough to keep them going until David or Bert can get some silage to them or, if the worst comes to the worst, get the sheep to the silage. It's the pigs which are Phil's main anxiety. If it goes on snowing is there enough grub up at Hollowtree to keep between 600 and 700 pigs going?

Promptly at 8 am, his regular starting time, Bert Fry turns up in the yard with his dog. One of the old school, it's a matter of pride for him to be there as usual. Over a cup of tea offered by Jill he explains that he saw it was snowing last night; he thought he might not be able to use the car so he got up earlier so that he could walk if necessary. You couldn't get through to the village with a vehicle, he assures them, but it's possible on foot if you dodge up into the field when you meet a drift.

With David clearing drifts from around the buildings with a bucket mounted on the tractor – a chore of doubtful value at the moment as it's still snowing – Bert is sent to help the cowman who is bound to be behindhand as every job takes twice as long in the snow. What Phil doesn't know at this stage is that Graham has had a series of hold-ups. First the automatic cluster removal system ceased to function, followed by the press button feeding. Just as he was getting used to managing without either of these facilities the parlour was plunged into darkness. What had happened was that snow had drifted into the parlour and lodged on top of various fuse boxes; as the heat from the cows had melted it, so the fuses blew.

After breakfast, Bert is despatched with a tractor and trailer bearing the milk, preceded by David with the tractor and bucket trying to clear the worst of the drifts for him. Once in the village he should be able to reach Wharton's Garage where he faces a long cold wait for the tanker. Phil asks David to get some hay to the outlying heifers. Susan Carter tells Phil over the telephone that Neil spent the night at Willow Farm and may still be there with his outdoor pigs. Then, wearing thigh boots and carrying a flask of coffee, he sets off across the fields towards Hollowtree.

Fortunately the crisis is short-lived on this occasion. The snow eases off. Brian's big 130 hp tractor with a mounted blade, on contract to the local Council, gets through later in the day and clears the way for the milk tanker and feed lorry (and, much to her vexation, enables Elizabeth to get to work the following day). But in The Bull, as they chew over events, extracting the maximum drama from each situation, there are the usual prophets of gloom who forecast worse to come. Phil has already rung the feed firm to have all his hoppers filled up. He remembers the last time they ran out, when he was making trips two or three times a week, bringing back a load in the Land Rover. The talk moves inevitably to the winter of 1947 when they were snowed up for 6 or 7 weeks, and when 18-year-old Phil Archer had to help his father dig the ewes out of the deep drifts. And of 1963 when no one could get their tractors started because the diesel had waxed up, and how the old petrol and paraffin Ferguson at Brookfield came into its own for a few days of glory. Out tumble all the cold weather stories, embellished for the occasion; of the frozen up milking parlour and how they set about milking 60 cows by hand until someone discovered how to unfreeze it, of how Fred Mason set fire to his barn using a blowlamp to thaw a water pipe and how the chap at Meadow Farm tried a similar remedy on a modern plastic water pipe – and melted it. They recall when it was so cold the eggs froze and split before they could be collected, and water went solid in the pails before they could carry it to thirsty stock. And then, just as the conversation is starting to flag, Bert Fry declares that in his opinion drought is worse than frost and snow, and off they go again.

The farming year may be said to start in October, but there's something about the birth of a new calendar year which tends to focus the mind on what lies ahead. On most farms it's a time for planning, making decisions, ordering supplies and budgeting. The scene of most of this cerebral activity is the farm office, accommodated at Brookfield in what used to be the old dairy, adjoining the house. A door opens from the yard into a hall out of which lead entrances to the kitchen on the left and the office on the right. A visitor to either of these rooms must first negotiate an assortment of articles which have been dumped there 'temporarily', some many moons ago. There's a torch, half a bag of fodder beet seed left over from last April, an axe handle, a drum of teat tip on its way to the parlour, a tin of wood preservative, a gas cylinder

Beef cattle often go through a market three times during their short lives; first as calves, later as growing cattle like these and finally as beasts ready for the butcher.

for the bird-scarer, a bag of dog food, Shula's old riding hat and several whips, Phil's shooting stick, two tins of paint, an umbrella and some draining rods; a box of dahlia tubers and some gardening gloves show that Jill is also guilty of adding to the collection.

The office itself is less confused than it looks at first if only because Rebecca, the travelling farm secretary, has put her foot down and restored some order. She comes half a day each week and needs to be able to find everything without much help from Phil and David. True there are piles of farming journals and cuttings, catalogues and reports all over the place, but the filing cabinet houses the essentials reasonably efficiently. A faded map of the farm and a calendar from the seeds firm hang on the wall beside the desk; David usually tries putting up the slightly less decorous offering from the machinery dealer behind the door, but since his mother has taken to using the office for her WRVS work it tends to get removed.

There are three good reasons for father and son to be in the office at this time of year. They can't get on with much outdoors, apart from the daily winter ritual. The sorting out and scheduling, the calculating and the telephoning are all just as essential to the operation of the farm as the feeding and mucking out; and, last but not least, it's warm in the office, with an endless supply of coffee from the kitchen next door. Every aspect of the farm comes under review. Should they invest in a new spreader, now that they've switched to buying their fertiliser in half-ton bags? Is it worth talking to Tony or Brian about sharing a forage harvester for silage-making? Are they lambing at the right time or would it pay to start earlier or later and what would be the knock-on effects on the rest of the farm? What would happen if they gave up growing oilseed rape? Rebecca could do some alternative budgets on her computer during her next visit. Each discussion is punctuated by activity on the calculator, by telephone calls to suppliers and other farmers who might help with advice, by combing through the advertisements in one of the farming papers or frantically searching for that pamphlet which was 'on the window-ledge the other day, I saw it'.

Most of their discussions are on-going but occasionally, over a cup of coffee in the warm office on a raw January morning, one or other of them will come up with something entirely new. David thinks they should consider breeding their own sows. Like most pig farmers these days they buy their replacements in as maiden gilts – young females which have not yet been put to the boar – from one of the firms specialising in breeding hybrid pigs. But David believes they could do the job themselves.

'I've been working out some figures Dad, and I reckon we could produce them at half the cost.'

'Half the cost?' Phil can't believe it. 'Why isn't everyone doing it?'

'Some of them are,' he replies. 'There's a chap I play rugger with – they've been breeding their own for a couple of years.'

Phil, with bitter memories of the last bill from the pig breeding firm, is very interested. 'But won't we lose hybrid vigour?'

'Not for some time – and if we do we can always go back for some more of their gilts.'

'We'd need to change the boars more regularly,' Phil ponders.

'Or use AI,' says his son. 'That's what my pal's doing. Think of the quality of the boars we could make use of then!'

'AI. Hmmm.' Phil digests the import of using artificial insemination on his sows. 'Have you spoken to Neil about this?'

'Not yet.'

'We'd have to bring him along with us on this one – if we do decide to do anything,' says Phil.

'According to this chap I've been talking to, it helps to cut down disease. You aren't bringing fresh bugs into the piggery and the pigs get used to the ones that are there.'

'Yes,' Phil's thinking hard. 'And it would mean we could select our own sows for breeding – you know, those nice long ones with 16 teats.' Most sows have only 14 teats and they both laugh comfortably at the significance of this in the knowledge that they may be on to a cost-saver. Then Phil has another thought. 'But wouldn't we need more buildings? Have you taken that into account, David?'

David reckons they could use some of the old buildings at Hollowtree but Phil reckons they'd have trouble mucking them out. And so the debate continues.

Meanwhile, up at the pig unit Neil is less concerned with hypothetical changes of policy than with feeding and mucking out. The sound of his car door slamming is enough to set off a cacophony of squealing from the older pigs. He has tried shutting it gently in the past but it only delays the reaction for a few moments – until he opens the door of the first house. He feeds the dry sows first because they make the most noise. These are the sows without piglets, most of them in some stage of pregnancy and they are housed in open buildings with yards. Along the front of the yards is a row of stalls and, as the sows rush in for their breakfast, Neil pulls a series of levers which lock each pig into an individual stall. This serves a double purpose; it prevents the sows fighting over the feed – they can become very vicious – and it enables Neil to clean the yard with the tractor later on without interference.

Next he checks that everything is all right with the sows and litters. There are 16 pens in the farrowing house and about three-quarters of them are occupied by sows and litters or sows waiting to farrow. The sows are confined in 'crates' which allow them reasonable freedom but prevent their lying down suddenly and squashing their piglets (200 kg of sow has a fatal effect on young piglets). Each pen has a heat lamp under which the little pigs lie when they are not suckling. Neil feeds each sow and casts a practised eye over the pen; years of experience tell him at a glance if anything is wrong. He looks carefully at a sow which is near pigging but decides it's not imminent.

The fatteners are still bawling their heads off. He wheels the feed trolley from pen to pen, silencing them 12 pigs at a time as he tips their nuts on to the floor. The first time she came to watch her husband at work Susan was horrified to see that there were no feeding troughs until he explained that pigs are very clean animals – they muck in a separate area which has a slatted floor so that the dung and urine fall though while the insulated floor of their living accommodation remains spotless. Now he trundles another feed trolley over to the growers, housed in strawed yards, and buckets the nuts into their trough, again looking for the tell-tale signs of trouble. Any pig which doesn't immediately come for its grub is suspect. By now a welcome quietness has settled over Hollowtree, broken only by the chomping of hundreds of jaws and the occasional squeal as pigs fight over the last mouthful.

Neil tops up the feed hoppers in the weaners' quarters; they're on ad lib feeding to get them off to a good start after being taken from their mothers at 3 weeks. (He moved two lots out yesterday and sterilised their pens as disease is never far away in a piggery.) Then he scrapes the yard, puts some more straw in their lying quarters, and lets the sows out of their stalls. He mucks out the sows in the farrowing house into a wheelbarrow and tosses some more straw in for the growers. Then he retires to what he calls his 'office' – a narrow cubicle in the entrance to the farrowing house where he keeps his precious records – and pours himself a cup of coffee from his flask while he decides what else there is to do.

HOME FARM

Before the last war a flock of 80 ewes would have been reckoned enough for one man to look after – and he would probably have pressed for a shepherd's boy to lend him a hand. Sammy Whipple tends 600 ewes single-handed, on a day-to-day basis, and has time to see to the beef cattle and help with the deer as well. Modern drugs which keep so many diseases and parasites at bay, extra help at peak times and bringing ewes indoors to lamb have all played their part in bringing this change about. Sammy would feel lost without his hypodermic syringe. He has a burly New Zealand student attached to him for lambing and he doesn't have to wander round the fields in the middle of the night looking for ewes in trouble, as his grandfather did; they're all in a well-lit shed.

Since Christmas, Sammy has been getting ready for the early lambing which begins about 5 January. He's been checking and re-checking his drugs and equipment – he keeps them in an old kitchen cabinet with a let-down front which he bought at a farm sale for £1 – and clearing out and refurbishing all the individual pens which he'll soon be needing. He has already sorted his ewes according to the colour of their raddle marks, so that he knows roughly in which order they'll lamb. Once they start things can become quite chaotic, although it tends to be a series of crises interspersed with periods of relative calm. There are 200 ewes due to lamb, most of them before the end of the month. If they were to lamb at regular intervals that would work out at one every 3 or 4 hours, in which case Sammy could probably manage without his New Zealander. But unfortunately they are not so obliging. One day only 5 ewes will lamb and another day more than 30. Sammy always reckons that most of them choose to lamb at night.

The ideal ewe would be one which lambed unaided, had healthy twins and enough milk to feed them, which 'cleansed' (parted with her afterbirth) naturally and took immediately to her lambs. Unfortunately some ewes have to be helped because the lambs are lying the wrong way or are too big for easy delivery, some have a dead lamb or lambs, some haven't enough milk, some retain their afterbirth which can cause serious complications, others refuse to take to one of their lambs, and a few die giving birth. The situation is not helped by having so many ewes lambing in a confined space. Left to herself, a ewe will always wander

Most ewes manage to lamb on their own but some need assistance especially when the lamb is very large or lying the wrong way. In this case the lamb is being born 'backwards'; it normally comes head first.

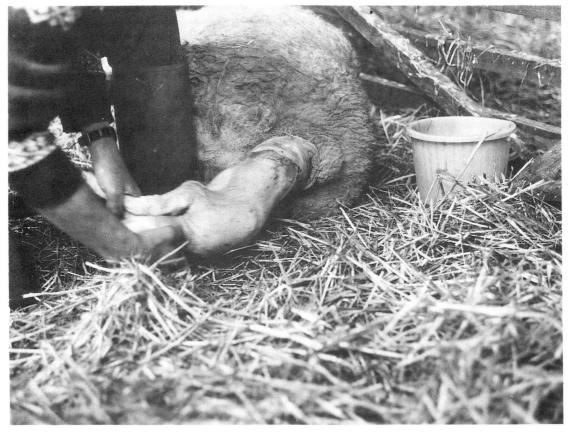

off on her own to lamb. Penned up in the lambing shed at Home Farm they can't get away from each other and 'mismothering' occurs; ewes 'claim' lambs which are not their own and in the hurly-burly of the lambing pen it's often difficult to sort out which one is which.

Brian usually does the late night or early morning shift – at least during the first fortnight when things are hectic. Sammy Whipple may knock off at about 7 pm telling Brian that there are a couple of ewes which look as if they could lamb in the next hour or two. Brian promises to keep an eye on them and in any case he'll go up to the shed at 11 pm. He finds on his arrival that all hell is let loose and he's forced to send Jenny for the already tired-out Sammy. By the time the shepherd arrives Brian has begun to sort out the situation but the noise is deafening. There's one ewe in difficulties, one dead lamb and one on its last legs, a ewe which hasn't lambed has 'adopted' one of the new-born lambs, one ewe is lying down, either having another lamb or in dire trouble. Tempers tend to get frayed in these situations and, against an earsplitting chorus of bleating, there's plenty of shouting between them as they try to sort out which lambs belong to which ewes. 'Look, Sammy, that one's hers.' 'No, this is hers, look,' as the ewe turns to nuzzle it. 'Then whose is this?' holding up a lamb by its forelegs. Sammy's too busy to answer. 'This one isn't going to last long unless we find the mother,' he shouts. 'I'm putting these two in there,' says Brian shoving a ewe into one of the individual pens and passing her lambs over. 'Have you squirted these?' (the lambs' navels must be treated to avoid infection). 'No, I haven't squirted any of them,' comes the reply over the clamour. 'Looks as if this ewe's had it.' 'I'll have a look at her in a moment,' yells Brian, 'What about this one?'

The frustration of trying to sort it out is matched only by the satisfaction when it's over with all the ewes in pens suckling their lambs. Some nights this just can't be achieved and they will go to bed, dog-tired, knowing that in the morning there's a ewe with no lambs or a couple of orphans still to be sorted out. They often make mistakes in the rough and tumble of the night before and there may be a sequel the following day. A self-satisfied Sammy may confront Brian with 'You know you put them two lambs in this pen? Well, one of 'em belonged in 'ere look,' and he shows Brian a ewe happily suckling twins. 'I thought they weren't quite right last night,' he ventures, seeing Brian smile.

Each ewe stays in an individual pen until it's clear that she is suckling her lambs properly – at least 24 hours. Then if the weather is reasonable she is turned out into one of the small fields near the house. If it's too cold or wet, she'll stay in an open yard with plenty of shelter until it improves. Brian likes to see every ewe go out with two lambs. As some have singles and some triplets – not to mention deaths – this involves a lot of patient matching up. By the end of January there will only be about 20 or 30 left to lamb. Everyone involved will be exceedingly tired and edgy and Sammy Whipple, who has worked about 160 hours overtime, will be considerably richer. Brian will be ready for his skiing holiday.

On a mainly arable farm like Brian's there's difficulty in finding enough work for the men during the winter months. They can pitch in during lambing to some extent, they can check over and repair the machinery and there is a certain amount of fencing and ditching to be done. But there is not really sufficient routine farmwork to keep them fully occupied. Brian is partly overcoming the situation by making sure that every time he takes on a new employee he acquires a new skill. Applicants are surprised to be asked whether they are accomplished carpenters, bricklayers or electricians. He has even taken on men with no farming experience at all. Brian's view – quite heretical in farming circles – is that it's much easier to teach someone to drive a tractor than to wire a building. This means that he is building up a team of craftsmen who, between them, can tackle all kinds of improvements; the beef building and the sheep housing were both erected and fitted out by farm labour, saving Brian a great deal of money as well as keeping his men busy.

Beef production is one of the least profitable enterprises on the farm and yet there seems to be no shortage of farmers prepared to keep beef cattle. This is because in many systems they 'fit in' and while they may not return as much as dairy cattle or even sheep they don't normally lose money. Brian justifies keeping them on several grounds. They eat up the by-products of his arable crops such as sugar beet tops and straw, which would otherwise go to waste, and turn them into beef and profit. In the summer he can use them to keep the grass in the deer paddocks under control; he couldn't use sheep as they have some diseases in common with deer. In the winter they turn a lot more of his straw into farmyard manure which helps him to maintain soil fertility. And, finally, he enjoys seeing them around and he's rich enough to be able to indulge himself even if they aren't very profitable. He would, however, be unlikely to keep them if they actually lost money.

Producing beef is not one of the most profitable of enterprises. Arable farmers, like Brian, often keep beef cattle to consume by-products such as beet tops or barley straw, and to produce valuable muck. These Hereford x Friesian heifers are eating silage.

BRIDGE FARM

During any frosty spell at this time of year, when he can get on to his land with a tractor without damaging it, Tony is busy carting and spreading farmyard manure. Livestock farmers will take any opportunity during the winter to get their muck out. The difference in Tony's case is that, as an organic farmer, it becomes a very positive act whereas in the case of many 'conventional' farmers it's more a matter of getting rid of it before it clogs up the system. To Tony, it's the main source of his soil fertility; he is not allowed to make up any deficiencies by spreading a few bags of compound or nitrogenous fertiliser an acre. Indeed the rules governing any farm which makes use of the Soil Association symbol to sell its organic produce are very strict. The Association's book of standards divides the various ways of maintaining fertility and controlling pests and diseases into four categories – 'recommended', 'permitted', 'restricted' and 'prohibited'. In the case of manure, Tony is 'recommended' to use only his own farmyard manure either composted or stockpiled under cover, though he's 'permitted' to use it either fresh from the cowhouse or stockpiled outdoors. 'Restricted' uses require the Association's permission and might include bought-in slurry or mushroom compost or sewage sludge. But Tony is totally 'prohibited' from using bought-in manure which isn't composted, or muck from systems regarded as unethical such as battery and broiler poultry units.

They were lucky at Bridge Farm to inherit a set-up known as 'loose housing' for their dairy cows from the previous tenant. Instead of the individual cubicles which they have at Brookfield, the cows lie all together on straw in a large building. It's not mucked out during the winter, so the straw and dung builds higher and higher. By the third week of April, when Tony reckons to turn the cows out, it will be 5 or 6 feet deep. The cows lie warm and comfortable, although the system does permit bullying and occasionally a cow gets a teat trodden on and torn. When the cows vacate the building he mucks it out with the tractor and stacks the muck somewhere near where it's going to be spread. It's one of these muck-heaps that Tony is dealing with this month if the weather allows. While stacked it will heat up and rot the straw, making it easy to handle with a mechanical spreader. So Tony

Muckspreading in midwinter. Intensively kept livestock generate a lot of manure and slurry which must be got on to the land whenever conditions allow. Tony Archer composts his and can be more choosy as to when he spreads it.

largely avoids the slurry nightmare faced by most dairy farmers. The slurry left in his collecting and dispersal yard at milking and around the self-feed silage clamp is scraped regularly and composted with straw to make yet more sweet-smelling solid farmyard manure.

Although he may not use conventional chemical fertilisers – the sort manufactured in factories – Tony is allowed to use certain naturally occurring minerals and other substances to supplement the manure. With plenty of livestock and most of his farm in clover-rich leys, Tony is in a favourable position to maintain soil fertility under an organic system. Meanwhile, he is gradually moving away from a position where he does not use chemical fertilisers because they are not allowed, to one where he avoids them because he no longer believes in them.

It's while Tony is spreading dung in Ashmead on the ground where he intends to plant his potatoes later on that he notices that the rabbits have been eating his crop of winter barley in Top Field. Since myxomatosis killed off about 90 per cent of the rabbits in the late fifties the population has ebbed and flowed according to the presence or absence of the disease. While Tony has been at Bridge Farm he has noticed how the rabbits will tend to build up in numbers to the point where they are doing him a lot of damage and then suddenly diminish due to a recurrence of the disease. Myxomatosis is spread by fleas, but how it is suddenly reintroduced into an area no one knows. Tom Forrest reckons that it's done deliberately by someone putting down a diseased rabbit among the healthy ones. What is certain is that it does not kill anything like as many rabbits as the original infection 30 years ago. So there are still rabbits about and they're making a mess of the headland round Tony's barley.

That night in The Bull he suggests to Eddie that he might like to bring his ferrets over on Saturday for a bit of sport.

'All right, mate,' says Eddie. 'I'll come over in the morning. They bolts better in the morning.'

'It's that boundary hedge, up the top of the farm.'

Eddie takes a swig of his beer. 'Ar, I knows. There's some really big buries there. And I knows just where to put me nets.'

'Yeah, and they aren't half pasting my corn,' Tony complains.

'OK, I'll come up Sat'dy, 'alf past ten. I'll bring Duke and Clarrie.'

'Clarrie?' Tony muses briefly. 'All right, I'll see if Pat wants to come.'

Eddie smirks, 'No, not that Clarrie. Clarrie's me other ferret, me jill.' He takes another drink and looks at Tony slyly. 'There's just one thing. Who's going to 'ave the rabbits?'

'I thought we'd go fifty-fifty.'

'Fifty-fifty?' Eddie explodes. 'Look, mate, I'm doing you a good turn. Stoppin' them beggars gnawing off your corn. Costs good money to keep ferrets.'

Tony knows when he's beaten. 'All right then – you can keep the rabbits, long as I have a brace for Pat.'

Eddie knows that he can sell the rabbits for nearly £2 a brace and if they're lucky they'll catch 20. Tony will have his reward in a morning's sport, a better crop of barley – and one of Pat's rabbit pies.

OVERLEAF Barn owl. Numbers badly affected by present-day farming, particularly through loss of suitable hunting habitat. Putting land into 'set-aside' could help a recovery.

TOP Middle White pigs were squeezed out 30–40 years ago and are now a 'rare breed'.

ABOVE A well-defined public footpath between two fields of rape. Carefully managed paths like this can harbour small mammals and even partridge.

OPPOSITE Although not always keen, most farmers will tolerate the hunt if it behaves well.

OPPOSITE Lesser whitethroat feeding its young. Removal of hedges and rough areas has threatened birds like this but sympathetic hedgerow management and replanting are now helping to increase numbers.

LEFT Sowing barley in good conditions with a pneumatic drill. The 'arm' guides the tractor on its next bout.

BELOW Ewes and lambs under cover. Shepherding is easier and losses reduced indoors.

BOTTOM The mallard – one of nature's survivors and often actively encouraged for sporting purposes.

ABOVE Interest in farm woodlands is growing with Government encouragement. Coppicing allows the same area to be 'harvested' regularly for stakes, hurdles and firewood.

RIGHT A craftsman at work. Economic pressures are forcing the replacement of the hedgelayer with the mechanical hedgetrimmer.

BELOW Emergency repair. Most farms now have someone who can carry out simple welding jobs.

OPPOSITE 'As hatched'. An old-fashioned flock with as many cocks as hens. Finding the eggs takes time!

TOP Most dairy farmers run a Hereford bull with their heifers as the most practical way of getting them in calf. After that they tend to use artificial insemination – from a Friesian bull for herd replacements or a Continental breed to provide beef calves.

ABOVE Although many ewes are kept inside during the winter this is largely for the benefit of the farmer. The sheep themselves, except in much more extreme conditions than these, are better off in the fields.

February

There's an old saying in Ambridge that winter's back breaks about the middle of February. But if the worst is over, as Phil Archer and his nephew Tony scrape the muck from their yards for the umpteenth time they know that it will be many weeks before the cows are turned out again. It's the time of year when farmers look anxiously at the hay barn and the silage clamp to satisfy themselves that they will have enough feed to last another two and a half months.

> Lock in your barn on Candlemas Day
> Half your corn and half your hay.

Candlemas Day falls on 2 February and what the ancient couplet is pointing out is that for the livestock farmer winter's only half-way through. This festival seems to have attracted more attention from the wise men who concoct country sayings than any other. So many of them dwell on the same subject that it's tempting to think that there must be some truth in them. They are all variations on the same theme. The weather at the beginning of February is apparently the key to what is to follow.

> If Candlemas Day be sunny and bright
> Winter will have another flight.
> But if it be dark and cloudy with rain
> Winter is gone and will not come again.

It's a teasing month, February. Snow and hard frosts give way to sunshine and a hint of spring before the weather relapses once more into unrelenting winter. For that reason it's an unpopular month with farmers who can never be sure how much, if anything, they'll be able to do in the fields. So Brian Aldridge flies back from Kitsbuhl by the middle of the month just in case it's dry enough to start on spring work – and most years wishes he'd carried on skiing for another week or two.

It's a busy time for gamekeepers like George Barford, who knows that there's nothing so fickle as a pheasant, and that if he doesn't keep on feeding his birds after the end of the shooting season they'll wander off to someone who does. Mr Woolley never understands this and assumes that once 1 February is over George is at a loose end and available for odd jobs. But in addition to garnering his birds he's busy trapping vermin and, later on, catching pheasants for breeding. Meanwhile the cock birds, safe until next shooting season, strut round Leader's Wood like favoured princes.

Baled silage, stored in airtight polythene bags, has become very popular in recent years for winter feeding. But if rats gnaw through the bags the silage quickly deteriorates.

Hunting continues throughout February as long as weather conditions permit, and woods put out of bounds during the shooting season are opened up to hounds once – so the cynics say – the keeper has made sure he's removed all the fox snares. A hound caught in one of these is a great giveway and undermines the supposed 'togetherness' of the hunting and shooting fraternities. Shooting has become such big business in some places that a keeper cannot allow anything to survive on or over his patch which may be a threat to the pheasants. So the stoats and weasels go, along with the rats and squirrels, the little owls, the kestrels and even the buzzards. It's true that the gnarled weather-beaten keeper, moving silently through the woods gun over arm, is a marvellous naturalist; he needs to be to know exactly where to place his traps.

February's a time when those impeccable squares of green velvet behind the houses in Glebelands are likely to receive an unwelcome visit from moles. No respecters of private property, these velvety creatures can wreak havoc on a lawn in one night, leaving a network of tunnels just below the surface together with the occasional molehill. They are equally a nuisance to farmers, killing off grass and encouraging weed growth with their hills, which later foul the mower or forage harvester. Thomas Tusser, that remarkable East Anglian 'Gent' recognised the trouble way back in the sixteenth century. Among the advice for the month of February in his *Five Hundred Points of Good Husbandry*, he writes:

Land meadow that yeerly is spared for hay
now fence it and spare it, and doong it ye may.
Get mowle catcher cunninglie mowle for to kill,
and harrow and cast abrode everie hill.

It's not so easy to find a mole catcher these days. Tom Forrest shoots any moles which threaten his little patch with his 12-bore shotgun, waiting for them early in the morning or after lunch, when they are likely to be working. He maintains that it's kinder than poisoning, the usual method on farms these days, and more certain than trapping.

Tom will be busy in his garden this month whenever he gets a chance, planting some broad beans, shallots and onion seed if it's dry enough, and a few lettuces under a cloche. As he wanders round he'll notice the rhubarb pushing its pink shoots through the soil. He'll see the buds of the elder in the hedge swelling and the catkins yellowing and getting longer. The birds are supposed to begin pairing on St Valentine's Day although it's usually later than this in Ambridge. But they'll be starting before the end of the month in most years. And Tom will soon be wandering down to the bridge over the Am to see if the dipper – that charming white-breasted water bird – is building a nest again behind the little waterfall where a stream joins the river.

BROOKFIELD

It's the first Tuesday in the month, the day the vet pays his routine visit. It's also the moment of truth for Phil since one of the jobs undertaken on these occasions is to check whether all the cows are actually in calf. Phil has been on a 'herd health scheme' with his vet for some time, one of the most important facets of which is to make sure that his cows are put back in calf on time; this involves Phil, the vet and the cowman, in keeping a close eye on all aspects of nutrition and disease which control a cow's ability to breed regularly.

The relationship between farmer and vet has changed in many respects since the days when Mr Farnon, the 'vetinry', held sway in *All Creatures Great and Small*, but in one way there has been little variation. The farmer still tries to make sure that he gets maximum value for his money, and the vet is still determined not to part with too much hard-earned knowledge for nothing. So when he arrives to do his PDs (pregnancy diagnoses) Phil has lined up several other non-urgent cases for him to have a look at. He knows that he is paying for the monthly visit by the hour, so he won't get any advice or treatment free of charge, but on the other hand it may save him having the vet out specially later on and so avoid a separate call-out fee. Ambridge farmers often joke among themselves that the vet seems to want £20 these days 'just to open the gate' – in other words simply for coming to the farm, before he starts examining anything. Phil doesn't begrudge the vet's fees, however, even though along with the drugs he supplies they come to several thousand pounds a year. He remembers that the last time the electrician came to see to Jill's dishwasher he seemed to charge more per

A tense moment for any dairy farmer as the vet checks whether his cows are pregnant. Failure to get a cow in calf at the right time can cost a farmer about £60.

hour than the vet – and Phil reckons that his cows are at least as important as his wife's dishes.

So this morning Phil has nearly 20 cows to be 'PD'd'; these are cows which calved last September and were inseminated in November. One by one they are put through the crush in the corner of the yard, and Phil and his cowman study the vet's face as he examines them internally. They know from past experience that a slight smile usually means that she's in calf; a worried look that she's not. It's the young vet, Martin Lambert today. They make a good team, he and Bill Robertson. Phil always feels that Bill has the edge in diagnosis – after all he has been at it for much longer – but that Martin may be a little better on treatment, slightly more up-to-date, perhaps.

Martin's brow furrows slightly as he examines cow number 414. 'How long did you say this one's been in calf?'

Phil looks at his notebook. 'Ten weeks, just over.'

Martin's still examining her. 'She's certainly not in calf that long. Are you sure you didn't see her bulling about a week ago?'

Phil and Graham exchange glances and then shake their heads.

'Well, keep an eye on her in a fortnight. If she doesn't come on heat then I'll have another look at her next month.' Martin removes his arm from inside the cow. 'Now, how many more are there?'

After the cows Phil presents his other cases, one by one, somewhat apologetically as if he'd only just thought of them although he, David and the cowman discussed them all thoroughly the previous day.

'Oh, while you're here, Martin, could you just have a look at this one – she trod on a nail? I gave her a jab but she's still very lame.' The vet examines the foot, asks Phil what he treated her with and goes to the boot of his car for an injection of a different antibiotic.

Next it's a heifer which Phil wants dehorned. Normally they disbud all the calves to prevent their horns growing but this one seems to have been missed. She already has horns 3 or 4 inches long and would tend to bully the other cows later on if they were not removed. He has to inject the heifer with an anaesthetic before he can saw the horns off, and while he's waiting for it to work Phil hauls him off to look at a calf which keeps blowing up. 'She's gone back down again now,' he says. 'Yesterday she was like a barrel, wasn't she, Graham?', he adds to convince Martin it isn't a false alarm. He's assured that it's a condition which will almost certainly right itself.

Much of the routine work which used to occupy so much of the veterinary surgeon's time is now undertaken by farmers themselves – the disbudding, the castration, the routine injections, even dealing with quite serious conditions like milk fever. The vets have had to come to terms with a certain amount of DIY treatment and are moving more and more in the direction of preventive medicine. More livestock is being taken to the surgery these days – especially sheep and calves – rather than incurring an expensive visit from the vet. And farmers like Phil, who maintain a dialogue with their vets, can obtain a lot of advice over the telephone. Phil keeps a supply of drugs on the farm and can give an ailing calf or lamb or pig a shot of antibiotic after chatting to the vet. 'If it's any worse in the morning give me another ring and I'll pop out,' he'll be told. In most cases it's better the next day.

Martin comes down to the farmhouse, has a wash and telephones the surgery. Phil decides that as he'll probably get charged for the coffee break anyway he may as well pick Martin's brains about a new treatment for mastitis he has read about. Then they go up to the pigs.

A prize for sheer audacity should be awarded to the genius who devised the term 'lagoon' for the giant cesspit used to store slurry on the modern dairy farm; it's the outstanding euphemism of all time. Anything further removed from the palm-fringed azure lake of the television advertisements it would be difficult to imagine. Phil's is a pond covering about a quarter of an acre adjoining the cows' yard in which is stored all the dung and urine of 95 cows. It's designed to hold the entire winter supply of slurry but no opportunity of lowering the level is ever ignored. It is only spread when it can be done without damaging the land; in other words when it is unseasonably dry or, more probably, when the ground is baked hard with frost. There are two good reasons why Phil likes to spread as much slurry as he can during the winter. Unlike Tony's farmyard manure which has been composted and is sweet-smelling, it taints the herbage, so it's better to apply it when the grass is dormant rather than when it's being grazed. And it smells quite strongly, especially the pig slurry from the tank at Hollowtree – so it's much less offensive in February than in June. They even managed to spread some near the barn conversions during a dry spell last winter without attracting complaints from the hyper-sensitive Mr Wendover. A tractor-drawn tanker is used to do the job; it sucks the liquid slurry from the lagoon under vacuum and then blows it out under pressure on to an inclined plate at the rear of the vehicle, which spreads it. Spreading slurry is not one of the most sought after jobs at Brookfield –

especially by Kenton who got sprayed with the stuff last year while unblocking the nozzle – but it's certainly one of the most essential to the successful running of the farm. Phil nearly got fined last spring when his slurry lagoon became too full and overflowed into the river Am, killing the fish.

Phil's ewes are due to start lambing around 25 February and it's amazing how accurate nature's clock can be. The gestation period for a ewe is 144 days and quite often the first lamb is born on time almost to the hour. Unlike Brian Aldridge, Phil does not have a building in which his flock spends the whole winter. He brings them into the barn a few days before lambing and turns them out again after they have lambed and settled to their offspring. During the 6 weeks prior to lambing they will have been fed a cereal mixture in addition to silage – starting at about ¼ lb per head per day and building up to 1½ lbs at lambing. The concentrates are fed in troughs and the person doing the feeding needs to be very agile as 300 hefty ewes converge on their breakfast. In spite of being warned about how violent they could be Mark Hebden volunteered to take the feed up one day, was literally hoisted off his feet by the voracious ewes and ended up in the mud. Shamefaced, he crept back to Brookfield where fortunately the first person he met was his mother-in-law who helped him to clean up a bit before he encountered the others.

By this stage of the winter at least half the Dutch barn will be empty of hay and straw, and a week or so before lambing David and Bert will be busy converting it into a maternity ward. Using a combination of spare gates, hurdles and straw bales they construct three large pens capable of holding 30 or 40 ewes apiece, and a number of individual pens to take the ewes once they have lambed. A week before lambing the flock is brought down to one of the fields adjoining the farm buildings and each day all the ewes which look to be close to lambing are brought into the barn or adjoining yards. By the end of February, lambing is in full swing.

There are days on the farm when everything seems to go wrong, so wrong it's almost as if Phil and David had sat up the night before and

A welcome sight to any shepherd as a ewe licks its newly-born lamb. An even better sight would be to find her fussing over twins – but perhaps she has another to come.

planned it. They're usually associated with or aggravated by bad weather, although oddly enough it's not normally the extreme weather which causes these 'bad days'; neither are the events in themselves all that dramatic. But this is the time of year when they often occur and this is the sort of thing which can happen.

It starts with a tap on the door at about 6.15 am. At first Phil doesn't hear it as he's listening to *Farming Today* on the radio. When he eventually opens the door he finds the cowman. 'There's a cow down, Mr Archer,' he says. 'I don't like the look of her at all. It doesn't look like milk fever to me. Do you think you ought to ring the vet?' Phil decides to have a look at the cow and follows Graham through the gloom over to the yard. On the way he notices that the yard is 1½ feet deep in water at one end; it must be a blocked drain. 'Been a rough night, Mr Archer. Did you hear that wind?' asks Graham. The cow is lying on her side in a pool of slurry in the dispersal yard. Phil doesn't like the look of her either. 'What's she doing in this yard?' he demands. 'Don't know,' comes the reply. 'Looks as if she was shut in here last night after milking, by mistake.' Phil frowns. 'Which means she hasn't had any silage all night. I wonder . . .' He bends over and smells the cow's breath. 'It's acetonaemia,' he declares. 'You have a niff.' (This is a condition caused by a shortage of glycogen in the animal's body. The breath has a characteristic sweet smell of pear drops.) Phil goes back to the farmhouse to ring the vet.

When he lifts the telephone he finds it dead. He doesn't have to wait long for an explanation. Bert arrives in the yard on foot. He hasn't been able to get his car in – there's a bough down across the drive by the road. 'Looks as if it's fouled the telephone wire as well, Mr Archer.' While Phil's working out the knock-on effects of that piece of information, Bert adds, 'There's a dead ewe by the gate in Trefoil. I spotted it in the car headlights.' So Phil can't ring the vet, or get the car out to ring from the village. The milk lorry will be coming before long and Phil's due at Court at 10 o'clock. David arrives on the scene at this stage and after being brought up-to-date on the situation throws in 'And don't forget we've got a load of ready-mixed concrete arriving this morning.'

Everything is resolved in the end but not before a further catalogue of frustrations. The chain saw refuses to start so they can't clear the offending branch. The milk lorry meantime arrives with a relief driver who refuses to hang about, so the milk isn't picked up and they have to use the emergency tank. Bert spends an hour looking for the rods to clear the drain in the yard before someone remembers that Eddie Grundy borrowed them a week ago and hasn't returned them. David offers to go across the field and down to the village to telephone – on the tractor. Unfortunately it won't go; water has got into the diesel fuel. Eventually Phil walks to the village and manages to summon the vet, postpone the concrete and ring British Telecom. It's only been light for an hour, and there's plenty of time for a lot more to go wrong.

HOME FARM

Brian is preparing to plant some more woodland, taking advantage of a new government scheme to encourage farmers to grow trees on good farm land. It's part of a move to take land out of agricultural production to avoid adding to the food surpluses, while at the same time improving the appearance of the countryside and producing something of which we import far too much – timber. Brian is less concerned with the motives behind the scheme than in improving his shoot with handsome grant-aid from the government. He's being paid about £500 an acre to plant the trees and nearly £80 an acre every year for thirty years while they're growing. He knows that he could make more than this by growing wheat or sugar beet, but he needs some more coverts for his pheasants and it pleases him to think that the Ministry of Agriculture is subsidising his shooting. He can boast in The Bull of doing his bit to avoid building more food mountains, while Mrs Snell continues to regard him as the conservation king of Ambridge.

So Brian, together with the woodlands specialist from ADAS who's advising on the project and Mike Tucker who will be doing the planting next month, pace thoughtfully round the corner of a field destined to be drilled with spring barley. They're deciding exactly how much land to devote to trees on this site and Mike is carrying a bundle of sticks to mark the boundary. In order to qualify for government help Brian has to agree to plant up at least 3 hectares – that's about 7½ acres – over a three-year period and the little wood they're planning at the moment is to be about 3 acres. He has selected the site carefully for two reasons. It's well placed in relation to other woodland on the farm; it's a couple of fields away from Leader's Wood, a comfortable distance for pheasants to fly when disturbed. And it fills an awkwardly shaped corner – one which has always been somewhat difficult to cultivate. They are arguing about the edge of the covert. Brian wants it to be straight to make it easier for machinery to operate, but the ADAS chap is suggesting rounding or angling it to make it more attractive. In the end they compromise and Mike bangs the stakes in to mark the line of the fence. It's to be a mixed wood, about 60 per cent hardwoods – mainly oak and ash, the local species, with some wild cherry and rowan – and the rest conifers, chiefly Norway spruce which produces good

Planting a new thorn hedge. For many years farmers were grant-aided to grub them and it is reckoned that a fifth of our hedges have been destroyed. Now there are grants for replanting.

roosting for pheasants. Mike will plant them in March, after the risk of hard frost, in large irregular-shaped groups to avoid regimentation, and protect them with plastic guards. He will also plant some shrubs to provide food and cover for the pheasants while the trees grow.

The wood will need maintenance for at least the next ten years and it will be much longer than that before Brian sees any return in terms of timber. There'll be some thinnings from the conifers in 25 years, but it will be nearer 50 before it yields the bulk, and hardwoods even longer. In the meantime Brian has added an appreciating asset to the farm and one which in time may well pay for itself in increased income from the shooting.

While on his way back from planning the new plantation, he notices that the wood pigeons are punishing his oilseed rape. As he opens the door of his Range Rover to inspect the damage a flock of several hundred get up with a flurry of wings, wheel round in the air, and alight on the crop about 200 yards away. 'Blast,' says Brian to the others, 'I'll have to get some of my little men up here after lunch. I've been wanting to try them out.'

'You know what you want Mr Aldridge,' says Mike, who rather fancies the job at £4 an hour. 'You want someone up here with a 12-bore.'

'I think I'll try my little men for a day or two,' Brian comes back. 'See how we get on.' Mike looks suitably mystified. A hundred acres of rape, which is a member of the cabbage family, must seem very tempting to a hungry pigeon in February. But a visitation can have quite devastating consequences, especially if followed by a hard frost and biting winds. The crop can be virtually killed off and need re-drilling in the spring, and apart from the cost of sowing again a spring-sown crop yields much less than one which stands the winter. So a farmer will try desperately to keep the pigeons off which, since the birds must eat, has the inevitable effect of driving them on to someone else's crops.

Brian sorts through his armoury of deterrents, the latest of which is a collection of bright orange 'men' made of plastic sheeting which alternately inflate and deflate, waving their arms in the process, to the accompaniment of ear-piercing shrieks. Brian hasn't tried them out before and, while hoping that they'll frighten the pigeons, is glad that the field is a long way from anyone's house. He remembers the trouble Phil Archer got into a few years ago with his bangers. 'Mind you,' Brian muses as he loads the gear into his vehicle, 'he really did ask for it, leaving them on all night, like that.' After planting the little 'men' in the field he goes on to check the ewes and lambs, now outdoors, before going to help Sammy feed the beef cattle and the deer which are still housed.

BRIDGE FARM

It's an inescapable fact of life for Tony and Pat that the worse the weather the better the prices for their produce; this is truer of organically grown vegetables because they are in shorter supply. They were watching an old film on television the other night in which pneumatic drills were being used to dig up leeks during the bad winter of 1947. Things have not deteriorated to that extent at Bridge Farm so far, but in order to cash in on higher returns from the organic co-op or local outlets Tony and Pat often have to operate in very unpleasant conditions. Some of the jobs such as riddling and bagging potatoes or sorting and packing carrots can be carried out under cover, although the barn on a freezing February morning can be pretty uncomfortable. But nothing in Pat and Tony's minds can compare with the agony of lifting leeks in bad weather. Time and time again they allow themselves, against their better judgement, to be tempted into braving the elements for an attractive price. The situation is nearly always the same; the co-op, or it may be another wholesaler, rings them up one morning to see if they can manage a few leeks.

'What, in this weather?' Tony starts off. 'You must be joking,' For once Pat and Tony are in full agreement – Pat is on the other side of the kitchen, sensing the enquiry, and shaking her head violently.

'The price is good,' comes the voice of the seducer.

'How much?' asks Tony tentatively.

Across the table, Pat whispers fiercely, 'Tony, no!'

'How does five quid a box grab you?' comes the answer.

'Five quid! Ouch.'

'That's what I sold some for this morning. They're getting desperate.'

Tony succumbs. 'I'll see what I can do.'

Meanwhile Pat clenches and unclenches her fists as she moans 'Oh, Tony, not again. Have you seen what it's like outside?'

But the decision has really been made. 'Five quid a box. That's what they're paying. Come on, love, we've got to lift some at that price.'

Pat tries every excuse – she's in the middle of making Tommy some pyjamas, the chap's coming to collect some yoghurt, she has promised to do some shopping for Mrs P, there's a hole in her gumboot – but it's no good, she knows she's trapped. Fifteen minutes later they're off up to the field with the tractor, a load of crates in the transport box and a fork to loosen the leeks as it's far too wet to do it mechanically. The temperature is hovering around freezing point and there's a persistent drizzle. Tony takes the tractor to the bottom end of the field and leaves it on the headland; the leeks are bigger in that area so he figures that they can lift the maximum weight in the minimum time.

Lifting leeks, tedious at any time and soul-destroying in bad weather. But it keeps up the cash flow over the winter. Often the worse the weather the better the price.

Tony digs and loosens the leeks. Pat lifts them, trims the worst of the roots and their clinging mud off with a knife and throws them into a crate. Water has collected between the leaves and the shaft of each leek, and frozen; the outsides of the leeks are slimy. They work in silence. Occasionally Pat will mutter, 'There's two or three there you haven't loosened properly', and Tony will go back and dig again. After a couple of hours of total misery, slipping and sliding about in the mud, Pat announces 'I've had enough. I'm going back.' Tony persuades her to fill another couple of crates and to promise to come out again after lunch. She agrees reluctantly, anything to earn respite from the torment.

By milking time they have lifted enough to interest the co-op, though fewer than Tony had hoped. He telephones to say he should have 20 boxes for collection the following day and comforts himself and Pat with the thought that they'll be worth nearly £100 at the weather-inflated price, about double the normal. But the job isn't over. Next morning he trims all the icy leaves and turns on the hose to wash them clean but finds it frozen up. He ends up washing the worst of the mud off in the sink in the back kitchen before packing them into boxes. Over a cup of coffee he and Pat wonder what else they could grow instead of leeks. 'There must be something,' declares Pat, but neither of them can think of another crop which would keep up the cash flow during the first 3 months of the year so effectively. They agree to keep on thinking.

Bridge Farm is infested with rats. It's nothing for Tony to be ashamed of; it can happen on any farm especially during the winter and where control measures have not been taken for some time. Farms are a haven for rats – there's always something to eat and plenty of shelter. It's not surprising that they make a bee-line for the buildings in the autumn when it starts to get colder and there's less food about in the fields. They can eat a surprising amount, but they spoil far more than they eat. And they spread infection, especially a very unpleasant condition know as Weil's disease. Neil Carter caught it at Brookfield in 1978 and was very ill as a result; it causes a form of jaundice and can prove fatal.

But it's not normally the danger of Weil's disease which first springs to a farmer's mind when he sees signs of rats, it's concern about the damage they are doing to his crops and feeding stuffs. Tony first noticed them when he was sorting through some potatoes and found some showing the signs of having been nibbled by rats. He then found droppings among his stored wheat and in the calf-house he found that they had been at the special mixture bagged up for the young calves; typically they had chewed the end out of not one but six sacks, with the result that there was about a hundredweight of feed spewed out on to the floor. As well as disease and waste, rats can also cause farm fires. The *Borchester Echo* carried a story of a farm towards Felpersham where they had gnawed through cables on a combine, the battery of which had been left on charge, and caused a full-scale fire.

Tony mentioned down at The Bull that he was laying poisoned bait for the rats. 'Pity about that,' Eddie said, 'I could have brought my ferrets up and we could have had a bit of sport.' To which Tony, remembering the last time Eddie came rabbiting at Bridge Farm, rejoined, 'Well, if you had, at least we wouldn't have argued about who was going to have the dead rats. You could have had the lot.'

March

It has been called the month of many weathers but the one element which seems to predominate is wind. Walter Gabriel's old granny always used to describe it as a lazy wind: 'Tis too idle to blow round you so it blows through you.' Before the days of tractor-cabs men would come in from the fields literally blue or purple with cold at this time of year, after 8 hours on a tractor seat.

Apart from its ability to chill to the bone – especially when it's in the east – March wind can have both a beneficial and an injurious effect. It can soon fan a flickering flame into a forest fire. More blazes are started inadvertently at this time of year than in the summer as everything is tinder dry and the sap has yet to rise. On the other hand nothing dries out the soil more effectively than a good wind; the frost has done its work over the winter in breaking down the clods and now the wind dries them ready to be broken into a seed bed. 'March dust to be sold, worth a ransom of gold', is one of many couplets and sayings about the value to the farmer of a dry March.

The early nesters like the thrush and blackbird are already laying their eggs while the crows and magpies hop along the leafless hedges searching for them – or their fledglings. Fortunately, the distraught mothers will lay again later in the month when there's more leaf to hide their nests. But it wasn't a thrush's egg or a blackbird's which Tom spotted in the beak of a magpie on a tree at the end of his garden, but a pigeon's. The magpie, with the egg forcing its beak wide apart, was dancing from branch to branch, followed in its every move by a crow. The crow fancied the egg for its elevenses; the magpie hadn't time to eat it and was determined not to lose it. The cavorting went on for several minutes before the crow edged the magpie on to the end of a branch from which the only escape was to fly off. In doing so it dropped the egg in some long grass. By this time the ever-practical Tom Forrest had picked up his gun, and as the crow wheeled round to see if it could spot the egg Tom shot it. The magpie, a bird practised at keeping just out of gunshot range, had disappeared. 'Well,' said Tom as he went back into the cottage, 'there'll be a few more birds on your table now.' Prue looked up enquiringly. 'I just shot a crow. He won't be robbing no more nests this spring.'

Fox-hunting tails off this month in Ambridge; some hunts carry on but there are too many lambs about in Borsetshire to make the hounds welcome visitors. Last March a vixen had her cubs in an old rabbit bury at the far end of the orchard at Bridge Farm. It's amazing how often they'll whelp near civilisation. 'Oh, Tony,' wailed Pat, 'she'll be sure to have my hens. You must drive her off before she does any harm.' But Tony brushed her anxiety aside. 'Don't worry. It's a well-known fact that a fox will never kill chickens on its own doorstep.' There is certainly evidence to support this piece of country lore, but it didn't stop the Bridge Farm vixen from taking three of Pat's hens in broad daylight a week later.

There's a wealth of early flowers for those prepared to search for them as they do their best to avoid the buffeting of the March winds. Wood anemones, sometimes known appropriately as the wind-flower, celandines, coltsfoot and red dead nettle are among the earliest, with violets and the first of the primroses not far behind. And in the Ambridge gardens the keen ones are planting seeds and giving their lawns a first trim perhaps. March is supposed to come in like a lion and go out like a lamb, but it's surprising how often the weather is actually kinder at the beginning of the month.

BROOKFIELD

The dominating activity throughout most of the month is lambing. There are lambs in the orchard, lambs in the adjoining fields, the various yards are full of lambs or ewes about to lamb, there are lambs in the loose boxes and in the kitchen and even, at times, in the Aga. Some farms practise a system of sheep farming known as 'minimal care', where in order to keep labour costs down to the lowest point a certain percentage of losses are expected and accepted. By the same token, at Brookfield it's 'maximal care'. Of course, they do incur losses but everything possible is done to save a lamb's life. This is partly due to a genuine farmer's concern and partly, it must be admitted, because Phil and David know that each lamb they can rear to 3 months or so is going to be worth about £40.

Every sickly lamb is put under the heat lamp or taken into the farm kitchen to increase its chances of survival. Some are fed at first with a tube which takes the colostrum straight into their stomachs. If a ewe has a single lamb or loses one of twins they will try to find her another – perhaps a triplet or one whose mother has died. Triplets are often left on the ewe which, although she only has two teats, may be able to feed all three. If she can't cope, Phil may try helping one lamb with a bottle of milk each day but leaving it with the ewe. Some ewes' milk is slow to come and the lambs may need hand-feeding until it does.

At times they have too many lambs and at other times too few to achieve their ideal of two lambs to each ewe. This is when they make use of the lamb bank run by the local NFU secretary. He keeps a register of farms with 'cade' lambs, and when a farmer rings up wanting a lamb he's able to give him one or two telephone numbers to try. Jill sometimes gets sent off in a hurry to bring a lamb back from a farm 4 or 5 miles away. 'And make certain it's a good strong one,' Phil shouts after her as she gets into her car. He knows from past experience that there's nothing worse than trying to put a weak lamb on to a ewe which doesn't want to take it. Far better to have one that's a week or two old and eager to suck.

The lambing shed is no place for the squeamish because out of 300 ewes lambing there are bound to be tragedies – lambs born dead or deformed, ewes dying or with prolapses – and a number have to be

A load of porkers on their way to the abattoir. The pig market is notoriously subject to ups and down. Too many pigs cause low prices which mean sows are slaughtered leading to fewer pigs and higher prices.

helped by David or Phil or Bert. They joke that the ewes prefer Ruth as a midwife because she has smaller hands. Sometimes a difficult birth is too complicated to be attempted on the farm and David or Phil will whisk a ewe off to the vet in the back of the Land Rover, often to have a Caesarian section. Since the abortion storm at Brookfield in 1987, anyone who is pregnant or suspects that they may be is deterred from going near the sheep at lambing. This is because the disease which causes enzootic abortion can also bring about miscarriages in human beings. It sometimes gives rise to knowing smiles and raised eyebrows when a female visitor to the farm refuses to come and have a look at the lambs – when usually it is simply because she hasn't time. Phil and Jill did think it was carrying caution a little too far, however, when Jennifer Aldridge cried off an invitation to *dinner* on the grounds that she was expecting a baby.

Sometimes, on his way to and from the lambing shed, Phil stops and has a 'chat' with Freda, his Middle White sow. These 'conversations' take the form of Phil saying something like, 'Well, how are you then, Freda? All right? I'll get you some more nuts next time I'm up this way', to which the sow replies with a stream of muffled grunts, the sure sign of a contented pig. Freda arrived at Brookfield unexpectedly a couple of years ago when Phil's old friend Jim Palmer had a stroke and needed someone to look after her until he was better. Sadly Mr Palmer died and his daughters, unable to cope with a sow, implored Phil to keep her. He gladly agreed having already become very fond of Freda, who in the meantime had produced a litter. With Ruth's help, he repaired an old pig ark and Freda was soon snuffling her way round the orchard, followed by her nine little piglets. Phil was often to be seen leaning over the gate watching them. David become quite concerned with his father's apparent obsession with the queer-looking and, as he saw it, outdated sow. What he failed to appreciate was that, in Freda, Phil was rediscovering some of the delights of farming which he had lost in the drive for efficiency. Of course he was proud of his hybrid sows up at Hollowtree and pleased when they had a run of good litters, but they didn't gladden his heart in the same way as the sight of Freda gently flopping on to her side under the old pear tree and, a moment or two later, nine fat piglets sucking away as she grunted peacefully. It was very clever to be able to wean a litter at 3 weeks, as they did at Hollowtree, but nothing like as satisfying as leaving Freda with her piglets until they were 8 or 9 weeks old and she was quite glad to get rid of them. It was all very well for Neil to boast that he had improved the feed conversion figure by 0.1 – but far more gratifying to chuck a bolted cabbage over the gate and watch the sow munch it.

David heaved a sigh of relief one day when he saw Phil drive off with the sow in the stock trailer. 'Thank heavens he's got over that,' David thought as he carried on stripping his digger. But 3 weeks later Freda was back; Phil had simply taken her to a Middle White boar.

Science has changed many farming practices since Phil was a boy, some of them fundamentally. But no boffin has yet overcome the necessity to milk a cow twice a day, 7 days a week. Phil doesn't do the milking regularly at Brookfield these days but he reckons to do it two or three times a week partly to give the cowman time off and partly to keep an eye on things.

On a bright warm day in June, when it's light soon after 4 o'clock, the birds are singing and the cows are lying out, it's quite a pleasant task. On a dark morning at the beginning of March, with a bitter east wind stinging his ears as he gropes his way across the yard, Phil sometimes wonders whether there isn't an easier way to earn a living.

Carrying his second mug of tea, he slops through the slurry in his gumboots as he looks up one row of cubicles after another, just to make sure that there's no trouble. The cows by this time are gathering by the gate to the collecting yard, showing no great sign of enthusiasm. Phil opens the gate and lets them in as he heads for the milking parlour. This is a highly functional blend of concrete and steel erected for the sole purpose of milking the cows as quickly and easily as possible. It's a herringbone parlour – so-called because the two lines of cows stand at an angle to each other as they're milked. The operator, in this case Phil, works from a pit in between them which conveniently puts his hands on the same level as their udders.

Having switched on the equipment, he allows the first eight cows in. Each one has a number freeze-branded on her hindquarters, and as they file into the parlour Phil taps the numbers on to a computerised panel at

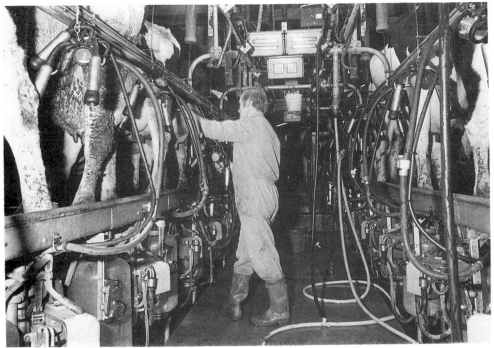

A modern herringbone parlour looks rather like the nerve centre of a space station. The central characters, the cows themselves, seem almost incidental to the technology.

the end of the pit. When all eight are in place, he presses a button and each one's ration of concentrates comes down a chute into the trough in front of her. Every cow is fed according to the amount of milk she's giving – or the amount she is felt capable of producing; the rations are programmed into the feeder once a month or sometimes more frequently.

Phil goes up the line hosing any dirty udders with warm water and wiping them with paper tissues. Before he slides his hand gently between the cow's hind legs to attach the teat cups he squirts a jet of milk from each quarter on to the parlour floor. He's looking for signs of the dreaded mastitis which shows up as clots in the milk. Then, with all eight clusters attached and the milk jetting into the glass jars alongside each cow, Phil lets the second eight in. They tend to enter in roughly the same order, day after day. If number 462 wasn't somewhere up in the first two lots, Phil would suspect that there was something wrong with her.

As each cow finishes milking the cluster is automatically pulled off her teats by a device operated by milk flow. This was one of the gadgets which David persuaded his father to install after he came back from his year on a Dutch dairy farm. Phil has to admit that it has removed a great deal of the strain from milking; if there's a sudden crisis and he has to go and sort it out it doesn't result in eight cows being overmilked, which can damage their udders. The pace of the actual milking is governed entirely by the cows; there's nothing Phil can do to speed it up. In fact it takes just about as long to milk a cow these days as it did at Brookfield forty years ago, before they had a milking machine – between 4 and 5 minutes. It's just that the machine milks them eight at a time. In the old days of hand-milking, about ten cows were reckoned to be as many as one milker could manage. Now one man is expected to look after well over 100 single-handed.

Before the first lot of cows is let out at the opposite end of the parlour, Phil squirts each of their udders with an iodine solution to ward off disease and protect their udders. Then, in come the next eight and so on until they are all milked. Phil doesn't like to hurry the job and takes nearly 2 hours to complete the milking; David can do it in under an hour and a half.

Half-way through David pops his head round the door at the head of the parlour to tell his father that he's going to top-dress the Croft, a field they hope to turn the cows on next month. A bag or two of nitrogen an acre will help to provide an early bite.

'Ah, David, just in time,' shouts Phil, across the gentle hiss and beat of the milking machines, '438's bulling. Can you hang on a minute and shunt her into the pen for me?' He wants her separated from the rest, ready for a visit from the AI man later in the day. David groans but stays; after all, he may be the one in the pit next time, with his father popping in opportunely.

It's not all straightforward. There's the cow that kicks the teat cups off before she's finished – usually, to Phil's annoyance, into a cowpat. And 403 likes to take a kick at the cowman. Being 3 feet down in the pit means that a kick from a cow can catch the operator where it hurts, and broken arms or bruised ribs are not unknown. It also places him in a vulnerable position when a cow decides to relieve itself, if he's caught unawares and hasn't moved out of range. A cow may be receiving antibiotics for mastitis or some other complaint, in which case Phil must remember not to let her milk go through with the rest. Occasionally there's a real crisis – like the time 421 suddenly pitched over with milk fever and fell into the pit. Graham, who was milking that day, nearly had half a ton of cow on top of him. It took the combined efforts of the vet, Phil, David, the cowman and Borchester Fire Brigade to get her out and into a loose box. Fortunately that sort of thing is not an everyday occurrence.

The milk all flows by pipeline into a vast bulk tank where it's cooled to 40° ready to be collected by the tanker. Four times a month, the tanker driver takes a sample, on the quality of which Phil's monthly cheque from the Milk Marketing Board is based. At this time of year most of the cows are past their peak, having calved in the autumn; they are averaging about 16 litres a day; (milk is sold in litres and then retailed in pints). Phil is sending away about 50 000 litres a month at this time of year and his March milk cheque will be more than £8000. In his peak month it can rise to well over £10 000.

HOME FARM

Like thousands of his fellow farmers up and down the country, Brian is top-dressing his winter-sown crops with a nitrogenous fertiliser. All crops need nitrogen, and although there is plenty of it occurring naturally in the soil it is not released quickly enough to satisfy the crops' requirements. Also, unlike the other main nutrients needed by plants – phosphate and potash – nitrogen is easily leached (washed away through the soil) and so has to be topped up at regular intervals if the crops are to achieve their potential yield. March is when most crops which have stood the winter need a boost and literally millions of tons of nitrogenous fertiliser will be applied before the month is out.

The response to nitrogen is dramatic; the experts say that up to the optimum every pound of nitrogen applied will give an extra 10 to 20 lbs of grain and that even with falling cereal prices it still pays very handsomely. After machinery and labour, fertiliser is Brian's biggest

Top dressing with a spinner. The use of chemical fertilisers, especially nitrogenous compounds, has increased enormously since the last war and is causing some concern.

input; he spends nearly £60 000 a year on it, much of it on straight nitrogen. Almost every day this month one of Brian's chaps will be spreading the stuff, first on the oilseed rape, then on the wheat and barley and finally on the grassland. The use of nitrogen on British farms has increased twentyfold since before the last war and some folk feel that farmers are using too much, especially in view of the proven link between nitrogen usage and increased yield and the cost of dealing with surpluses.

Thoughts about restricting the use of nitrogen are running through the mind of Robert Snell as, early for an appointment in Felpersham, he stops to watch Brian's tractor driver at work in a huge field of winter wheat. Near the gate stands a trailer loaded with fertiliser in half-ton bags, while the tractor threads its way up and down the 'tramlines' spinning the nitrogen on to the crop which stands 6 to 9 inches high. As concern over the side-effects of high-tech farming battle in Robert's mind with admiration for its achievements, Brian pulls up in his Range Rover. He has dropped by to see whether they will need more fertiliser sent up to finish the field. They lean over the gate, watching them load the spinner.

'Does it ever worry you, all this nitrogen?' asks Robert Snell affably.

'Worry me? Good God no. Why should it worry me?' Brian laughs cheerfully.

'Well, I read somewhere how you farmers keep on using more and more.'

Brian snorts. 'Well, look at that wheat. All yellow. If ever a crop needed a shot in the arm it's this one. If I didn't top-dress it now we'd get a lousy crop. Wouldn't even cover our costs.'

'And what about that field?' Robert points at the other side of the road. 'Will that have some nitrogen, as well?'

'Too true,' says Brian. 'I'm applying 1½ hundredweight an acre all round – and another couple of hundredweight the end of next month.'

Robert winces and Brian continues enthusiastically. 'And some more just before it comes into ear. I want 4 ton an acre from this wheat; shan't get it but that's what I'd like – and you don't get that without plenty of nitrogen.'

'But I thought . . .'

'I know what you're going to say,' Brian cuts in. 'We don't need all that wheat. Look, it's not up to me to solve that one. I'm running a business and the more nitrogen I use, within reason, the more I stand to make.'

'How about if everyone used less?' Robert persisted. 'If nitrogen was limited.'

'Wouldn't work,' Brian looked serious. 'You couldn't have quotas. Too complicated. Couldn't police it.'

'But how about if they *taxed* nitrogen? I believe there are comparatively few factories in the Community. They could tax it at source. Wouldn't that stop farmers using as much?'

Brian broke in quickly, 'Wouldn't stop me using as much.'

'Not if they put a really huge tax on?'

'It they put a huge tax on, it would shove the price of food up and who's going to like that?' says Brian, beginning to lose his patience.

But Robert plods on. 'The price of food may go up a bit but think what would be saved – all those billions they spend on storing surpluses. They could subsidise the food out of that, and they could use the money from the tax on nitrogen to help farmers in other ways. They would save all that fossil fuel making nitrogen and we wouldn't have these nitrates in the water. It all makes sense, you know.'

Brian's getting back into his Range Rover. 'Look, I've got to go. They're going to be out of fertiliser here before long. Must send someone up with some more. Cheerio.' And he drives off.

Although highly effective in increasing yields, nitrogen has brought its own difficulties. Applied heavily to cereals it can make the straw tall and weak so that it is prone to 'lodging' – in rain or wind it can collapse before it is ripe, making harvesting very difficult and reducing yield considerably. However, the agro-chemists have not been idle and nowadays the farmer who wants to use the maximum application of nitrogen also sprays his wheat with a growth regulator which shortens and stiffens the straw. Nitrogen has the effect of darkening the foliage of crops to which it has been applied; sometimes, in May, Brian's wheat will look almost blue with nitrogen; yet the plants are sturdy and he hopes that the fertiliser is benefiting the grain rather than the leaf and stem.

There has been general concern in recent years at the level of nitrate in our water supplies. Nitrate is a compound of nitrogen and oxygen and a vital requirement of plants. An increasing amount of nitrate is seeping into aquifers and rivers and hence into drinking water. No one knows for certain what harm this is causing, but the content in some parts of the country, mainly in the eastern half but extending west in places almost to the Welsh Borders, has risen to an unacceptable level. Farmers agree that their increased use of nitrogen is partly to blame for this and a major drive is under way to try and reduce nitrate levels which are several times the European Commission permitted figure. Farmers are asked to apply nitrogen only when crops are actively growing so that it is all used up and not leached out and, as far as possible, not to leave land uncropped so that naturally occurring nitrate is utilised and not available for leaching.

March is a busy month at Home Farm with Brian trying, but not always succeeding, to drill at least some of his sugar beet, peas and spring barley. If the soil is too wet to get a seed bed – and sugar beet is

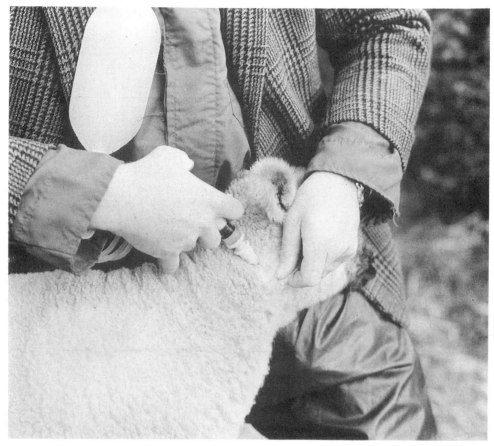

Modern drugs administered by the shepherd have made it possible for one man to look after huge flocks of sheep singlehanded. A vaccine given a fortnight before lambing wards off eight diseases likely to affect ewe or lamb.

particularly fussy in this respect – or to operate drills, there is nothing he can do about it except wait.

Brian starts lambing his second batch of ewes – 400 of them – in the last week of the month, and a fortnight before this he injects each ewe with a vaccine which protects them and their lambs against eight diseases. The injection is given in the neck to avoid damaging the carcass; the lambs derive protection subsequently from the milk. Sheep suffer from a wide range of diseases, some with highly colourful names.

Among those controlled by the vaccine are pulpy kidney, braxy, blackleg and struck. Then there are black disease and blackquarter, bighead and louping ill, scrapie and contagious pustular dermatitis (also known as orf, from which Jennifer suffered some years ago after helping with the lambing). Gid or sturdy is caused by a tapeworm, liver fluke is spread by snails, and husk brought about by lungworms. Sheep can become infested with ticks, lice, mites, keds and maggots, and among the functional disorders a shepherd must be on the look-out for are pine, swayback, double scalp, yellowses, rape scald – and dafties. Most, but not all, will react to modern drugs or other treatment if discovered in time. Once lambing begins it brings the same panic as in January; the difference is that, with better weather and the grass growing, it's easier to turn the ewes out once they've lambed.

Red deer, looking a bit 'moth-eaten' after a winter in the yard but otherwise in good fettle. Present profitability is largely dependent on the sale of breeding stock.

BRIDGE FARM

When Tony and Pat became fully organic in 1988 it meant changes in the way in which they managed their livestock. No more monthly deliveries of ready-compounded dairy nuts from the feeding stuffs firm; no more routine use of antibiotics to control diseases. The bulk of the feed for the livestock now comes from their own organically produced fodder, and the approach to animal health is based on prevention of disease rather than treatment.

As with the use of chemical fertilisers and sprays on crops, the Soil Association lays down firm standards for any organic farmer who wishes to use its symbol to market produce. They begin by ruling out 'unethical' systems of keeping stock, such as permanent housing of breeding stock or prolonged confining or tethering. The rules governing feeding have been drawn up in the knowledge that at present there just are not enough organically produced feeding stuffs to satisfy requirements. So Tony is allowed to include in the ration for his cows a maximum of 20 per cent of feed from 'conventional' sources – ie grown with the help of chemicals. When there is enough organically produced feeding stuff to go round, the rules will be changed to exclude non-organic feeds totally.

In practical terms at Bridge Farm, this means that for dairy rations Tony first uses the barley he has grown himself, and he supplements this with barley or wheat from other farms bearing the Soil Association symbol. To provide the protein necessary for milk production and general well-being he buys field beans or linseed cake, not necessarily from organic farms (since less of this constituent is required) and he adds some seaweed powder to provide essential minerals and trace elements. So once a month, instead of the lorry from the feed firm rumbling into his yard it's the mobile mill which grinds and mixes the ingredients to produce a palatable meal with a protein content of 14 or 15 per cent. In addition to this, of course, the cows are fed silage made from his own organically managed swards during the winter, and throughout the summer they graze the same fields. Organic farmers believe that sound nutrition is the basis of the health and vitality of their livestock.

When it comes to controlling disease the organic philosophy is that health is not simply the absence of disease but also the ability to resist infection and other disorders. Most modern dairy farmers rely heavily on the use of drugs, especially in the control of mastitis, the ever-present threat. The Soil Association standards allow conventional drugs to be used only to save life or prevent unnecessary suffering, or to treat a condition where no other effective treatment is locally available. The conventional treatment for a cow with mastitis is to squeeze a couple of tubes of antibiotic up the affected teat and keep the milk out of the bulk tank for the next few days (tests for traces of antibiotics are made regularly at the dairy). Tony and Pat now cure most of their cases without drugs – by hosing down the udder and frequent milking. It doesn't always work. In his early days of organic farming Tony persisted for too long with a particularly vicious mastitis bug and nearly lost the cow as a result. It was only when the vet told him if he didn't use an antibiotic the cow would die that he relented. Occasionally Tony is tempted to slip back into the old well-tried ways, but Pat usually manages to bring him up short. 'When you think,' she remarked recently, 'that the average dairy cow in this country doesn't even complete its third lactation – it's not a very good advertisement for antibiotics, is it?' So Tony keeps hosing his cows' udders in the hope that they will achieve the ten or twelve lactations that he knows are possible.

Meanwhile Pat and Tony are gently feeling their way into homoeopathic treatment of their stock. Since there are very few veterinary surgeons in the country who specialise in homoeopathy this involves a fair amount of do-it-yourself trial and error. They have already discovered that it doesn't have the sure-fire curative powers of antibiotics but, encouraged by Pat, they are determined to persevere. As with human homoeopathy, the principle is to treat diseases with small quantities of drugs that excite symptoms similar to those of the disease or, to put it more simply, curing a condition by giving 'a hair of the dog that bit you'.

Early in March the postman will deliver an expected but unwelcome letter to Bridge Farm. It's a bill for £3500 from the estate office, being the half-year's rent due on Lady Day, 25 March. Rent Day has changed a great deal since Squire Lawson-Hope sat behind his huge desk with the agent at his side, while the tenants in their shiny blue serge suits shuffled in one after the other to pay their dues and exchange somewhat one-sided pleasantries with their landlord. Tony simply pops round in the Land Rover in his working clothes and hands over a cheque – to his cousin Shula Hebden, from Rodway and Watson who manage the estate on behalf of his sister Lilian Bellamy – or his mother, Mrs Peggy Archer, who works part-time in the office. Few estates can enjoy such close family relationships. On estates which still employ a resident agent, rent audit can still be quite a formal occasion with the tenants looking smart in their market clothes, joking in a restrained manner with one another and each hoping for the opportunity of a few words with the agent which might result in some improvement of their holdings.

Britain's farmers are divided into owner-occupiers like Phil Archer and Brian Aldridge who own their land, and tenant-farmers like Tony and Joe Grundy who rent the land they farm. The landlord-tenant system has had a very long and successful history, although the amount of tenanted land has diminished sharply since the beginning of this century. In 1900 about 90 per cent of England was rented; now it is only a third of that. Tenants are much better protected from landlords these days than at one time when they could be turned out of their farms at will. Except for those who have entered into tenancy agreements since 1984 all agricultural tenants enjoy security for three generations, although those wishing to succeed must fulfil certain qualifications. For example, if Eddie Grundy wishes to succeed his father at Grange Farm he has to satisfy the landlord that, in addition to being a close relative, he has obtained a living from the farm for five out of the last seven years, he is healthy, he is financially sound and that he has the right training and experience to farm the land properly. And one day William may be entitled to succeed to the tenancy on the same terms.

Although lower now than earlier in the 1980s, the price of farm land has risen spectacularly over the last 30 years or so. Land which changed hands for £100 an acre in 1960 may now be worth £2000 an acre, and as a result many farmers have become very rich – at least on paper. Brian Aldridge is worth several million, mainly due to inflation in land prices; and most of the Brookfield land was bought in the 1960s at a fraction of its current value which would make Phil Archer very comfortably off if he decided to sell up. Had Tony been able to buy Bridge Farm in 1978 instead of renting it, he would now be worth about £140 000 instead of £40 000 odd, even allowing for the necessary mortgage. But it's not as simple as that. The plain fact is that he could not have afforded to buy it; the farm was not capable of making enough money to cover the mortgage repayments. This is because the price of the land, on which the mortgage is based, is out of proportion to its earning capacity, while rents are much more in line. The Phils and Brians of this world must count themselves lucky that they owned their farms during the massive rise in land prices. Tony must content himself with having obtained the tenancy of a farm with the right to pass it on to his son and grandson.

The tenancy agreement signed when taking over a farm is quite an intimidating document designed mainly to safeguard the landlord's interests. It binds the tenant to look after the land, keep it in good heart, maintain the hedges, ditches and fences and control the weeds. It reserves many rights for the landlord, including the timber and the game. Tony is not allowed to cut down a tree, without permission, or to shoot a pheasant on Bridge Farm. The tenant may gather stones from his fields to make roads and he may shoot rabbits, but most other activities are specifically excluded. These days, tenancy agreements tend to be interpreted more liberally than at one time, partly because modern farming methods have made slavish adherence to some requirements less important – weeds and diseases, for example, can now be controlled by chemicals rather than by crop rotations. But a tenant must still suffer the landlord's friends or sporting syndicate tramping and driving round his fields on shooting days in search of the pheasants which he knows have been feeding on his corn.

In spite of the disadvantages of renting a farm, however, tenanted holdings are still hard to come by. Tony appreciates this; he also knows that the one certain way of being given notice to quit is failing to pay the rent. So, painful though he finds it, he once again takes his cheque round to the estate office on Lady Day.

April

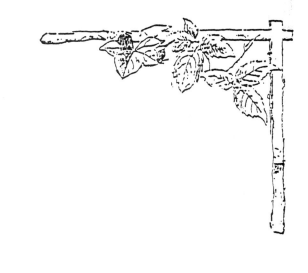

'Sweet April showers doo spring maie flowers' according to Thomas Tusser, the sixteenth-century writer, but the Ambridge farmers are less concerned with encouraging next month's blossom than with the miraculous effect traditional April weather has on their young crops. Although it can be marred with cold winds, the occasional frost and sometimes even snow, April rarely fails, at some stage, to provide the magical conditions so important to germination and early growth. Gentle rain followed by sun followed by more rain and more sun suits Brian's sugar beet, Phil's spring barley and Tony's organic wheat equally. It also makes the grass grow.

The hedges begin to green up again this month, giving vital cover for nesting birds, and the trees put on their spring clothes – although it's surprising how many varieties in Ambridge are still leafless by the end of the month, the oak and ash often among them. Meanwhile the early fruit trees are a mass of blossom; in the orchard at Brookfield it's usually the delicate damson flowers which come first, followed soon afterwards by the more robust pear and cherry.

That annual phenomenon, the arrival of the cuckoo, takes place around 16 or 17 April. If anyone in Ambridge hears it much earlier than that it's almost certain to be Eddie Grundy up to his tricks again. Shortly afterwards the swallows reappear, following their awe-inspiring flight from Africa, to nest in exactly the same buildings as last year. The Bridge Farm children had some fun one very dry April when the swallows were having difficulty finding enough mud to rebuild their nests – obviously flying long distances to collect it. Pat suggested to John that he fetch a bucket of wet mud from the gateway down the drive and tip it in the corner of the yard where they could see it from the kitchen window. John didn't see why Helen couldn't fetch it but eventually they did it together and were rewarded by a non-stop procession of swallows picking up bits of mud and swooping into the calfhouse.

The winter programme of evening activities tends to come to an end in the countryside as the days lengthen and farmers and gardeners get busy outdoors again. This month sees the local point-to-point races. These are organised by the hunt partly for enjoyment but mainly these days to raise funds to meet the mounting expenses of hunting. All the horses taking part in the races have had to qualify by hunting during the season although many come from other hunts. The event attracts people from Borchester, Felpersham and even Birmingham – the more the merrier from the point of view of the organisers who charge £10 a car for the best site near the bookmakers and refreshment tents, regardless of its occupants, and £7 further afield.

Out come the baskets of drinks and the picnic lunches and it's generally an occasion enjoyed by many who have never followed hounds in their lives and never will. Farmers, who provide the most important ingredient of hunting – the land – are admitted free of charge although these days instead of receiving a complimentary pass for the car they are asked to pay at the gate and reclaim the fee from the secretary. Few of them do, although David Archer always threatens to, and so the hunt benefits still further. The truth is that most farmers who enjoy hunting are happy to contribute; those who don't approve of the sport are unlikely to attend the point-to-point anyway.

BROOKFIELD

Half-way through the farming year, which began in October, finds Phil Archer with the satisfaction of the autumn and the frustrations of the winter behind him, and the promise and challenge of the spring and summer lying ahead. Although at the beginning of the month his cows are still indoors day and night, the end of the 6-month daily ritual of feeding, bedding and slurry-scraping is in sight. For the last few weeks he has been spinning nitrogen on to his grassland to encourage an 'early bite'. Lambing is all but over with only a few stragglers left, and the ewes will be in the fields with their lambs but still receiving a few nuts to keep them milking until there is plenty of grass. This is the time of year when the young lambs, bursting with exuberance, race each other round the field. Like children, they drift away from their mothers and then suddenly, as if a starting flag has been dropped, they're off. After two or three bouts they return to the ewes and start sucking furiously, their tails waggling with sheer delight. At Hollowtree, the pigs continue

'The time of year when young lambs, bursting with exuberance, race each other round the field.'

to be born, weaned and fattened oblivious of the fact that it's spring and the sticky buds on the horse chestnut outside The Bull are bursting into leaf once more.

In the fields, the winter-sown wheat and barley and the oilseed rape will have greened up again nicely after the winter frost and winds. Phil usually manages to drill his spring barley about the first or second week of April, according to the weather. This will be sown on land which he wasn't able to drill last autumn or perhaps a grass field which he has just ploughed up. Around the middle of the month he will drill his fodder beet and plant his potatoes, and that completes his cropping for this year.

The date for turning the cows out, rather like the decision as to when to bring them in in the autumn, is something which is not fixed but tends to 'emerge' in the light of weather conditions and grass growth. There could be plenty of grass but the ground might be so wet that Phil knows

Once the cows are turned out in the spring they find the bulk of their own feed and spread their own muck which reduces the cowman's daily routine.

that the cows would poach it – tread it up with their feet – so that a lot of the precious grass would be wasted. On the other hand the land could be dry enough but the grass hasn't responded to the top-dressing sufficiently well, usually because it has been too cold. In either of these cases the cows would be kept in a week or two longer, assuming that there was still silage in the clamp. The worst mistake a dairy farmer can make in the spring is to turn the cows out before there's enough grass to keep them going, perhaps because the silage is running short.

As in the autumn, the cows themselves take part in the decision as to when they should go out. They seem to be able to sense when the conditions are right and keep walking over to the gate in their yard, bellowing peevishly. Dan Archer used to have a rule of thumb learned from his father. 'When your foot can cover five daisies,' he used to say, 'it's time to put the cows out.' David used to tease him about its depending on whether one took size 8 or size 13 boots, but the simple fact is that there aren't really enough daisies around now for the test to be applied effectively. However, like all the old saws there was something in it; the growth and flowering of the daisy would say a lot about the season and, presumably, the growth of grass.

When it comes to the appointed time, they usually start at Brookfield by letting the cows out between milkings, keeping them in at night for the first week. Once they're out all the time there's not much slurry scraping to do – but instead they have to be fetched in twice a day for milking and taken back afterwards. This is easy when they're in an adjoining field but more difficult when they're in one of the fields up the road. The cowman usually borrows the Land Rover or the old tractor to save his feet, and the cows soon adapt to a new situation.

Around the middle of the month it's Phil's turn to use the potato planter belonging to his local farmers' group. The group has been updating its equipment, and for the first time it's a fully automatic machine which arrives, requiring only a driver. The one it replaced needed people on the back to place the potatoes in cups on a revolving wheel which then buried them beneath the ridges. The locals gather at Brookfield to watch the new technology which includes a device which 'sieves' all the stones and clods from the ridges, depositing them in between the rows where they can't do any harm. Instead of the seed being taken out to the fields

An automatic potato planter at work. Control of the crop is exercised by the Potato Marketing Board in an attempt to balance production and consumption. If there are too many the surplus goes for stockfeed and the grower is compensated.

in bags it now comes in half-ton boxes which are tipped into the hopper of the planter from a forklift truck. Tom Forrest shakes his head in wonderment as the machine sets off across the field, leaving the ridges of fine tilth with the seed buried at the correct spacings. He asks Phil what has happened to Maud, Lil, Madge and Connie, the gang of women who used to accompany the machine. Would they be back in the autumn with the harvester to tease Neil and Bert and anyone else in the team? Phil shakes his head. It seems that the manned harvester is to be replaced too; with no stones and clods left in the ridges there won't be anything for the women to pick out. 'Ah, well,' says Tom, goggling at the new machine. 'You can't stand in the way of progress.'

Unlike some crops, the growing of potatoes in this country has been controlled for more than 50 years, although there is pressure from some quarters to return to a 'free market'. Phil has been allocated a basic acreage by the Potato Marketing Board – in his case 15 acres. Each year the Board works out how many potatoes are needed and prescribes a 'quota' expressed as a percentage of each grower's basic acreage. Random inspections are carried out to make sure that quotas are not being exceeded; sometimes these are carried out from the air. Growers are allowed to sell only to licensed merchants and the size and quality of the tubers is regulated. The working of the Potato Board is financed mainly by levies on growers – Phil has to pay about £30 for each acre of potatoes he grows.

The object of what might at first appear to be excessive control is to try and equate home production of potatoes with home demand. Although some potatoes are imported – mainly out-of-season earlies and those for processing – more than 80 per cent of the potatoes eaten here are home grown. Phil is one of some 20 000 growers producing between 6 and 7 million tons a year; his acreage is a bit smaller than average. Even with regulated growing and marketing, gluts and shortages still occur because no one can control the weather.

The price of trying to equate production and consumption is self-discipline on the part of growers and an element of price maintenance in the shops. No one controls the retail price, but it is obvious that if there are just enough potatoes prices will not fluctuate wildly. The alternative is a free-for-all in which growers would stand to make a lot of profit some years and lose money in others as acreages and output varied. Some of the larger specialist growers favour this but chaps like Phil prefer stability. Some years they'll do better than others, depending on the weather, but as long as they make a reasonable job of growing them they are not likely to make a loss on a flooded market. Potato consumption in this country has been going up in recent years, and there are fears that if consumers found that the 50p bag of spuds had gone up to £1 because of a shortage, they might go home with a packet of rice or pasta instead.

HOME FARM

Every tractor is busy at the beginning of the month as Brian endeavours to keep up to date with his work. Priority is given to drilling the sugar beet, but he has plenty of other important jobs to do. There are peas and spring-sown barley to be got in, top-dressing of grassland and cereals with nitrogenous fertiliser, spraying of cereals to kill weeds, harrowing of grassland to tear out all the dead herbage, rolling of cereals and grass. Brian is rarely to be seen on a tractor himself, except to keep the drill going through lunchtime if the weather looks uncertain. He's the oiler of wheels, the one who tries to make sure that there are no hold-ups. This can involve him in all sorts of things, from delivering a tractor driver's flask of coffee to the fields to avoid his having to go back to fetch it, to taking a worker's wife into Borchester to the ante-natal clinic, to save her husband having to take a few hours off at a busy time.

Lambing is in full swing for the first half of April, declining after that but with some stragglers still to lamb at the end of the month. Taking care to choose a warm period, Sammy Whipple will turn out the beef cattle; they are quite sensitive to the cold and can easily develop a kind of pneumonia if put out too early. A number of the cattle will have been sold for slaughter during the winter and Brian will build up the numbers by buying at market, this month or next. If he has enough grass to feed them on he'll buy early, as the price tends to grow with the grass.

Drilling sugar beet into a good seedbed. Advances in seed production, precision drilling and chemical weed control have all but eliminated the back-breaking hoeing, once a feature of the crop.

Spraying winter barley to control weeds. With falling margins on cereal growing farmers are now carrying out less routine spraying with expensive chemicals.

Also, when the conditions are favourable, Brian turns out his deer. They have, like the cattle, been housed since November, fed on silage and a cereal mixture and although in good condition, seem as if they could do with a change of scene. Their coats look a bit moth-eaten at this time of year, but a few weeks on grass will work wonders. Before they are loosed into their paddocks they are dosed with a drug to control worms, one of the deer's greatest enemies. Brian is building up his herd of red deer and now has nearly 100 hinds, due to start calving at the end of May, and 55 young deer of both sexes born last summer.

One of the features of deer farming which sometimes alarms the onlooker, as well as the farmers themselves, is the forbiddingly tall fencing. It evokes visions of concentration camps or at least safari parks – you wouldn't be surprised to see guard towers at intervals – and suggests that the inmates are, to say the least, extremely dangerous. This is the last reaction on earth desired by those involved, who wish to see deer accepted on farms as naturally as cattle or sheep; they hope that as

In recent years much of the profit on producing fat lambs has come in the form of subsidy. There have been times when the taxpayers' contribution almost equalled the price paid at market.

numbers increase deer farming will lose its unorthodox image. It's still in its infancy in this country (although not in New Zealand where 10 per cent of farmers are involved); the first commercial deer farm was not established here until the early seventies. Only about one farmer in a thousand keeps deer and there are still fewer than 20 000 hinds on British farms (this compares with about 18 million breeding ewes and 1.3 million beef cows).

The deer paddocks at Home Farm are carefully situated close to the buildings so that Brian can keep an eye on them. It's land which is rougher and more undulating than most of the farm, with a few trees and bushes to give shelter, and some damp patches ideal for 'wallows'. Deer, like pigs, adore wallowing in mud, but that is the only similarity. In many ways they behave like sheep, although they weigh much more than ewes (a hind weighs 2 to 2½ cwt compared with an average ewe at about 1½ cwt). At present they are much more profitable than either sheep or cattle because most of the female offspring are sold to other farmers for breeding, leaving only the males to be slaughtered for venison. If they all had to go for venison at current prices deer would make less profit than well-farmed beef cattle or sheep. This is what worries Brian in the long term, once the market for breeding hinds is over. But his friends in the British Deer Farming Association tell him there's so much scope for expansion that he'll be drawing his old age pension long before that day arrives.

So Brian keeps his fingers crossed and continues to expand his own deer enterprise, wishing at the same time that Sammy Whipple was a bit keener on them.

In April Brian begins marketing the first of his lambs born in January. These days lambs are only about 3 months old when they are slaughtered, to meet the demand for leaner carcasses and smaller joints. The price is high at this time of the year and goes down steadily from now on as more lambs become available. Only farmers like Brian who have gone to the trouble and expense of lambing early are able to cash in on the higher prices. So he's keen to get his lambs to market at the earliest possible opportunity. Later in the season he leaves it to Sammy Whipple, but for the first few weeks he wants to be there when they are selected.

Behind the sheep building at Home Farm is a modern system of handling pens where they can drench, dip, shear and sort the ewes and lambs with the minimum of effort. First they separate the lambs from the ewes (known as 'shedding') by driving them down a race with a gate at the end which can turn them into one of two pens. Inevitably there's some shouting as the odd ewe is let in with the lambs and vice versa. Then Sammy begins to 'draw' the lambs which he thinks are ready for the butcher, with plenty of verbal help from Brian who stands outside the pen, not wanting to get his trousers covered in grease from the wool.

'There, behind you,' he shouts, above the noise of ewes and lambs. Sheep tend to baa only when separated ewe from lamb, but they certainly make up for their silence at other times on these occasions.

Sammy grabs a lamb. 'No, no,' bellows Brian. 'Against the rails.'

Sammy tries another, but gets it wrong again. 'With the speckledy face,' bawls Brian. 'That's the one,' he cries as Sammy holds the lamb against the side of the pen with his leg and quickly feels its back. 'No,' yells Sammy. 'He's too light.'

And so it goes on. Each time Sammy finds one which he estimates is the correct weight and carrying the right amount of meat he manhandles it over to the gate and Brian lets it into the next pen. Each man knows what he is looking for, but lambs are deceiving creatures and there are plenty of false alarms before both are satisfied that they have drawn all the likely ones. The next job is to weigh them individually – they are aiming for 40 kg, but at this time of the year will settle for a couple of kilograms less. Brian marks each one he wants to sell on the head, but before they load them up he hisses, 'Let those others out, Sammy. Can't hear yourself think here.' He knows that once the lambs are back with the ewes the din will abate.

At Borchester Market the lambs have to be weighed again – this time for subsidy purposes. Under the EC Sheepmeat Scheme, the farmer is paid a premium if the price falls below a set of guide prices fixed by the Commission. This time Brian is pleased with his first 20 lambs. They sell well and will bring him well over £1000 – with the taxpayer's contribution.

BRIDGE FARM

Tony is engaged on a job which occupies many hours of his time during the year – mending his equipment. On this occasion it's his corn drill with which he's hoping to plant some spring barley. Like much of his machinery it was purchased at a farm sale for a fraction of the cost of a new one and has seen better days. David sometimes teases him about what he describes as his 'museum' of old farm tools, but he argues with some justification that there's no point in having a machine costing thousands of pounds, which he uses only a few days a year, rusting away when he could buy a second-hand one for a few hundred and invest the rest in something which would make money, such as another cow. David contends that he could be doing something to make money during all the time he spends on repairs but Tony sticks to his guns.

April is a comfortably busy month at Bridge Farm. There's wheat and often barley to be sown and later the potatoes to plant; Tony usually persuades Phil to send the group's planter round once it has finished at Brookfield. The rest of his crops are not put in until May or June, but he's getting the land ready for them. For most of the month the cows are still indoors in their winter routine which Tony feels, as he scrapes the slurry for the 150th time, has gone on long enough. However, because of his organic system he can't 'tickle' the pasture up with a spot of nitrogen, and so his grass is slower to grow in the spring; the nitrogen nodules on the clover roots, which will later provide him with free nitrate, don't get to work until the soil has warmed up. So it's usually towards the end of the month when his cows go out for good. He can't help feeling a little envious when he sees Brookfield's cows strip-grazing a nice ley when his are still coming into the parlour with dung-coated udders, but he consoles himself by thinking of how much he has saved by not using nitrogen.

The 'in' word in agricultural circles these days is diversification. Farmers have been so successful at their job that almost everything they produce is in surplus. In order to reduce output, support prices from the EC are being cut and farmers are being encouraged to earn part of their living from non-agricultural enterprises. Instead of bolstering prices for crops and livestock the government is now offering grants to farmers to develop businesses which are either extensions of conventional food production or completely new ventures.

Tony and Pat have already responded by starting to process their own organic milk into things like yoghurt, cream, butter and soft cheese. Adding value to produce before it leaves the farm is one of the prime recommendations of the Ministry of Agriculture.

Farmers are fortunate in having a number of built-in benefits to hand when it comes to diversifying. They have land, often woods, sometimes water, all of which can be exploited for the leisure market. They have buildings, frequently surplus to agricultural requirements, which can be adapted to house alternative enterprises; and they possess certain skills needed to run a business. These advantages put them in a strong position to develop other sources of income as returns from their traditional output become less rewarding. Tony is not in as favourable a situation as some, being a tenant farmer. His landlord raised no objection to his building the dairy, but might not be as keen if he wanted to open a farm shop or start building pine furniture in the barn.

There have been some remarkable examples of diversification in Borsetshire in recent years. A farmer near Edgeley, for example, has turned one of his woods into a battle ground – enthusiasts come from miles away at weekends to indulge in mock war games, firing capsules of paint at one another instead of real bullets. He reckons that it pays far better than wheat or potatoes. In the south of the county a farmer has created a nature and farm trail which attracts visitors from as far away as Birmingham, including a lot of school parties. At Loxley Barratt a chap has flooded a low-lying part of his farm, stocked it with trout and let the fishing to an engineering firm in Felpersham at a figure which leaves him a healthy margin after paying for establishment and running costs.

Tony is convinced that he could make a success of all the things he sees other farmers doing – if only his landlord and the planners would allow him. Pat, more sensibly, suggests that going organic in itself is a form of diversification and that cashing in on the healthy demand for organic produce is what they should be doing. She regards the yoghurt as only the beginning. It cropped up one evening recently when Tony was thumbing through the *Borchester Echo*.

'Hey, listen to this, Pat. There's a bloke the other side of Borchester getting a grant for starting up some livery stables in his old cowhouse.'

'Good luck to him, I say,' came the reply.

'Yes, but Pat. We could do that here, he's charging £50 a week for each horse!'

Pat wearily looked up from sewing a button on John's school shirt. 'Who's going to look after them? We don't know anything about horses.'

'Helen does. She'll be leaving school soon.'

'Tony, she's only nine,' sighed Pat. 'Anyway, aren't we better off sticking to what we know? We haven't begun to exhaust the organic thing yet. Why don't we have a go at organic beef or lamb? Or even pork? We could grow more barley. I'm sure there's a market.'

But Tony is lost again in the *Echo*. 'Six horses he's going to take to start with,' he muttered. 'That's £300 a week. Blimey.'

May

No other month in the calendar has beguiled the lyricist over the years so much as 'the merry month of May'. 'Dewy May', 'bursting boughs of May', 'bright May morning', 'the darling buds of May' and other eulogies abound. Among them is the enigmatic children's song 'Here we go gathering nuts in May', enigmatic inasmuch as no tree in Ambridge produces its fruit at this time of year, and certainly the voracious squirrels will not have left any of last autumn's harvest. Perhaps it meant the decorative oak apples which local children used to pick on 29 May (Oak Apple Day, which commemorates the restoration of Charles II).

The rest of the children's song continues 'on a cold and frosty morning', which reminds us that in spite of all the panegyrics 'May's new-fangled mirth' sometimes has its downside. It can be cold and wet – and frosty as Tom Forrest well remembers, having lost more than one row of early potatoes in his time to a sharp May frost. Arable farmers like a wet May which brings the corn on well, although the silage-makers prefer it to dry off in the second half of the month when they are busy filling their clamps.

Whatever the weather, May is one of the richest months of the year, overflowing with bird song and wild flowers. Even

Brian's hedgerows are coming back to life now that he has stopped spraying the headlands and hedge bottoms. But as someone was pointing out in The Bull the other evening, it's not only spraying which kills the wild flowers. Fertilisers can have the same effect over a longer period as can too much grazing; primroses don't survive long when the land is stocked heavily with sheep. There are still cowslips to be found round Ambridge. Prue Forrest knows where they are but she no longer picks them to make wine. Instead she gathers may or hawthorn flowers which grow in profusion, and makes a simple wine of crystal clarity.

At the beginning of the month the verges are yellow with dandelions, a beautiful flower if only it restricted itself to verges. Unfortunately, one of Joe Grundy's new leys was also a mass of dandelions last May. He was wondering how he could get rid of them, before they took over totally from the grass and clover, when he spotted a couple picking the flowers into baskets. He watched them with growing incredulity and then decided to investigate.

By the time he reached them his boots were yellow with pollen. 'What do you think you're doing, then?' he asked.

The answer was self-evident. 'Picking dandelions.'

Joe's mind worked overtime. Was there some hitherto unperceived commercial value in the blooms? 'What you going to do with them?' he demanded suspiciously.

'Make wine,' came the answer. Joe's brow furrowed as he tried to find some way of turning the situation to his advantage. Could he and Eddie make dandelion wine in the barn, he wondered – like cider? He'd have to think about that one. So he resorted to a holding operation. 'Wine, eh? Well, they must be worth *something* to you.'

BROOKFIELD

The most important job at Brookfield in May is making the silage, although they don't usually start much before the third week. The early part of the month is taken up with preparations such as cleaning out the clamps, putting the sides on the trailers and overhauling the forage harvester. These tasks are punctuated by frequent trips to the fields to see how the grass is coming on and anxious consultations between Phil

Silage making is about the most hectic time of year for the dairy farmer. The grass is cut earlier and wilted to improve quality before being picked up and chopped by the forage harvester and carted to the clamp.

and David. Since his year on an intensive grassland farm in Holland, David has become the acknowledged expert on grass at Brookfield. He's the one who decides things like how much nitrogen to apply and when to cut for silage.

Silage is the mainstay of the winter feeding at Brookfield for both cattle and sheep. Although the technique of conserving grass in this way is at least a century old, it's only since the last war that it has become progressively more and more popular to the point where haymaking has been pushed firmly into second place. Two advantages which silage has over hay are that it's not so dependent on the weather for successful conservation, and that the cattle can eat it direct from the clamp without the need for carting.

Grass pickled in its own vinegar is how silage has been described. The sugars in the grass are fermented by bacteria to produce acids which preserve it. Normally when a pile of grass is left, the lawn mowings for example, it quickly heats up and goes rotten. In making silage the bacterial action is controlled by excluding the air, so that only the 'right' bacteria are allowed to operate. This is done by consolidating the grass as it's tipped on to the clamp, and by keeping it covered with a plastic sheet afterwards. To watch the frenetic activity taking place at Brookfield or any other farm during silage making, an observer would never guess at the complexity of the process involved, but there is a crucial balance between a number of factors including the state of maturity of the grass, its dry matter content and the length to which it is chopped by the forage harvester. If the Archers get it all right they are rewarded with a pleasant-smelling, palatable product which the cows will love and milk well on. If they get it wrong they end up with foul-smelling silage on which the stock won't thrive very well. A lot of the uncertainty of silage making can be avoided by using an additive to aid fermentation – usually either sugar or acid or sometimes bacteria – which is applied to the grass before it goes into the clamp.

Some time during May Phil will have to shear his sheep, or rather have them shorn, since these days this work is usually carried out by gangs who travel from farm to farm. Occasionally they may be New Zealand students having a working holiday in the UK, but normally in Borsetshire they are young farmers who get together at this time of year to earn themselves a little extra money. They will usually adapt their services to meet the farmer's requirements; he may have enough labour to manage everything except the actual shearing, but the gang will often supply a complete team including someone to catch the ewes and someone to roll the fleeces. Charges average about 50p a ewe depending on how much is undertaken.

In some parts of the world, like Australia, sheep are kept primarily for their wool and the meat is almost a by-product. In this country it's the other way round and the wool comes as a 'bonus'. Its sale is organised by the British Wool Marketing Board, which is responsible for handling the clip from 90 000 farmers. Between them they keep 35 million sheep, 20 million of which are shorn each year (the rest being lambs which haven't grown a full fleece). The total clip is nearly 40 million kg. Farmers complain that the price of wool hasn't kept pace with other farm products; Phil would receive perhaps £750 for the fleeces from his 300 ewes, out of which he pays £150 to have them

Weaners in a straw yard. Some pig farmers specialise in producing weaners and others in the fattening process but at Hollowtree they breed and finish their own pigs. It takes about six months from birth to bacon.

shorn, leaving £600 (about the same as he'd expect for a dozen lambs). Meanwhile Jill can never understand why the wool price is so low when she sees what they ask for a jumper in Underwoods.

The date of shearing is fixed by a combination of factors including the weather, the availability of the shearers and other commitments such as silage making. The warmer the weather the earlier Phil would like to shear. On the one hand warm weather increases the chance of maggots. (These are caused by blowflies which lay their eggs mainly on the mucky wool around the tail; when they hatch out they start feeding on the flesh of the sheep, and removing the wool usually prevents this trouble.) But warm weather also induces the 'rise' of grease into the wool which makes shearing much easier and increases the weight of the fleece. So if it weren't for the threat of maggots Phil would probably leave shearing until the beginning of June. However, he prefers not to have to worry while silage making, so he tends to have them shorn before they start on silage.

Before the shearers come the ewes will have been dagged, that is the mucky wool will have been clipped from their backsides. Not a popular job this, but one which Jethro Larkin always took in his stride with good humour and which Bert Fry now seems to have taken on. This deters the flies and also ensures that dung-encrusted wool is not sent off to the wool merchant where it will be down-graded and result in a lower price. Tom Forrest is always keen to know when Phil will be doing this job as he likes to scrounge some daggings and put a bucketful in a tank of water for the tomatoes and other selected crops in his garden.

Shearing day tends to be busy and noisy but good-humoured. It usually takes most of two days unless the men can start first thing in the morning. But they work fast; indeed Jethro used to get very upset at the speed at which they worked, and had difficulty keeping up with them when he was wrapping the fleeces and stowing them in the huge sacks. He wasn't used to piece work. One year they brought their own wrapper, a slip of a girl who found plenty of time in between coping with the fleeces to tease the shearers and Jethro who was catching ewes for them that year.

Jill often comes out while they're shearing, with tea and cakes. She loves to see the lambs searching for their clipped mothers; it doesn't take them long and they're soon lifting them off the ground as they enjoy a good suckle. The ewes have lost 4 or 5 lbs of wool apiece and look unexpectedly white against their lambs as they wander off together; the lambs are not shorn. Farming is becoming a more solitary occupation with most jobs carried out by a single operator from the tractor cab, so everyone seems to enjoy the few opportunities for teamwork which evoke memories of times not so long past.

During Phil's working life the farm labour force has shrunk dramatically. It's not possible to make direct comparisons on the basis of Brookfield because the farm is nearly five times as big as it was 40 years ago, but the regular workforce in the country as a whole is not much more than a quarter of its post-war size. Whether the workers were attracted away from the land by the promise of higher wages, or displaced by more and more machinery, the situation now is that there are more farmers than farmworkers. The labour pattern emerging is that as far as possible a farmer likes to arrange things so that he can do most of the work with family labour. Where this is not possible he will tend to look for a contractor who supplies not only the machine he can't justify himself but also the operator, representing invaluable extra labour. Tony Archer can manage his silage making only because Gerry Goodway brings the forage harvester – and himself. Phil's potato planting is made much easier because a driver comes with the machine. Farming has always been seasonal but modern methods have made it even more so. Silage making, which now provides for most of the winter feeding, lasts about 10 days; when Phil was a lad they would have been haymaking at Brookfield on and off for perhaps 6 weeks. Shearing can be completed in a day whereas at one time it would last a fortnight. So there's a temptation to look for casual labour to handle the peaks, and manage without at other times. Phil hires Neil Carter as a self-employed man, although it's on a regular part-time basis at Hollowtree. If a farmer has to take on someone permanently, he's inclined to look for a student like Ruth or a YTS lad like Steve at a good deal less than the basic wage of a farmworker.

When Phil decided to take on Bert Fry he was going very much against the trend. The truth was that he missed the reliability of Jethro who was killed in an accident in 1987. Ruth Pritchard (as she then was) had been a successful pre-college student, but as Phil entered his sixties and with David developing other interests he simply wanted another Jethro, someone who'd live in Woodbine Cottage, who'd know what to do if he or David weren't there and who'd be as interested in Brookfield as he was himself. Bert Fry, who had been foreman on a farm not unlike Brookfield, seemed just right. He was delighted to find another job at 56, even though it meant a lower wage. (He had been earning a basic £130 a week as a foreman, but was entitled only to £120 as a general farmworker with a craftsman's certificate.) In addition he was offered a rent-free cottage, a quart of milk a day and overtime at £4.50 an hour. He seemed content enough with the conditions, and accepted the job.

Most sheep shearing these days is done by gangs – sometimes of New Zealanders – who travel from farm to farm. In this country farmers keep sheep mainly for meat and regard the wool very much as a by-product.

HOME FARM

The early part of May sees the Ambridge countryside slashed with huge swathes of chrome yellow. This is the blossoming of the oilseed rape, a crop virtually unknown until less than 20 years ago. It was actually grown here 300 years ago, but was ousted by cheap imports in the nineteenth century. In 1970 a mere 10 000 acres were grown; 17 years later this had increased by nearly a hundredfold to close on a million acres. Part of this incredible rise was due to the attractive prices guaranteed to growers once we joined the EC, and partly because it is a crop which fits very well into the arable rotation. The enormous expansion in cereal growing in this country since the last war has brought with it an upsurge in pests, diseases and weeds associated with corn crops. Oilseed rape, a close relative of the cabbage, is what's known as a 'break crop' – it can be grown to break the sequence of cereal crops and the troubles which accompany them, since most of these are specific to cereals. But it has the additional advantage, unlike other break crops such as sugar beet or potatoes, that it can be drilled and harvested with the same equipment as wheat and barley.

Brian Aldridge was the first farmer in the Ambridge area to grow oilseed rape but others, including Phil Archer, were not slow in following. In the early seventies there were remarkably few pests and diseases, but now there seems to be a bug for every stage of growth including the cabbage stem flea beetle, the pollen beetle, the seed weevil and the pod midge. Most of the rape is sown in the autumn, soon after harvest. Brian usually drills his about the end of August. It grows to a height of 5 or 6 feet and after its dramatic flowering sets seeds, like a cabbage plant, in long thin pods. The crop is grown for its tiny black seed (about 100 000 to the pound) which is crushed to produce oil. Some of this is used industrially for lubrication – at one time, being very slow-burning, rapeseed oil was used in lamps on trains – but most of it nowadays ends up in margarine or cooking oil. The crushed seed is used as a high protein constituent of animal feeding stuffs. The price has become less attractive in recent years which has led both Brian and Phil to experiment with other possible break crops – Brian has been trying peas and Phil field beans. But yellow fields will continue to be a feature of the May landscape in Ambridge for some time to come.

May is not a particularly busy month for Brian; all his crops are drilled and growing. Like Phil he will be making silage in the second half of the month and he'll also have a visit from the shearers. The main task of the month is spraying against pests and diseases. Brian will be involved in deciding which crop is to be sprayed with what chemical and when. Spraying is being taken increasingly seriously by the authorities, and safety standards have been tightened considerably in recent years to avoid danger, either to the farmer and his staff or to the consumer of products which have been sprayed. It's ironic that as pesticides are tending to become less toxic the handling of them on farms has become more carefully controlled.

Pesticides are mainly of three types: herbicides for killing weeds,

Spraying sugar beet to control mildew. Safety standards have been tightened considerably in recent years to avoid danger to the operator or the consumer of sprayed crops.

fungicides for controlling diseases and insecticides for destroying insects. In few branches of agricultural science has more skill and ingenuity been displayed than in the development of these chemicals.

At the end of May some farmers go round the edges of their cornfields and spray a narrow strip with a chemical which kills everything. The idea is to create a kind of *cordon sanitaire* around the crop and prevent weeds creeping in from the hedgerow. Brian doesn't do this but follows a practice recommended by the Game Conservancy which involves leaving a wide strip round the headland unsprayed by harmful chemicals. The idea of this is to encourage the partridge. It was not that the chemicals were poisoning the birds; they were killing off all the tiny insects and weeds on which they fed. Brian began this practice in the mid-eighties and has been rewarded with a marked increase in the partridge population.

Brian's hinds begin calving about 25 May, and for a week beforehand he is keeping a close watch on them through his field glasses; the less disturbance they have at this period the better. Most of them will calve on their own, but once calving has started Brian will go round the paddocks twice a day to ear-tag the calves. The hind gives birth to her calf in some long grass or nettles away from the rest of the herd, and the young calf remains there, with its mother visiting it four or five times a day to feed it. Brian's only chance of getting near the calf is when it's very young. He wants to tag and sex it and also note the number of its mother so that they can be related subsequently.

The hind usually objects to interference with her offspring and in rare cases will even reject them after tagging, leaving Brian the time-consuming job of bringing them up on a bottle – or persuading someone else to, as he did with Betty a couple of years ago. Sometimes the hind will attack, and Brian always carries a sack ready to throw over her head to distract her. The calves are often difficult to locate when new-born, and taking a car or tractor into the paddock to look for them may be dangerous. Brian has found that the best vehicle is his newly acquired Japanese four-wheeled 'motor bike' which makes it easy for him to cover the ground while giving him excellent all-round vision.

BRIDGE FARM

In early May Tony prepares his land for carrots. Since he is not permitted to control weeds with chemicals the preparation is more involved than it would be on a conventional farm. A fortnight before he intends to sow the crop he works the ground down to a fine tilth and then ridges it. This exposes a greater surface of soil in which weed seeds can germinate. By the time he drills the seed – towards the end of the month – the ridges are covered in weed seedlings. He borrows a precision drill, which places the seed at 3 cm spacings, from another organic farmer with whom he has formed a friendship.

Carrots usually take about a fortnight to come through but in very favourable circumstances can make it in just over a week. The main method of weed control in organic carrots consists of burning off all the young weed growth just before the carrots emerge. Normally this is carried out with a flame-weeder about 10 days after drilling, but obviously the longer Tony can leave it before the carrot seedlings come through the more weeds he can destroy and the less need there will be for subsequent hand weeding. A couple of years ago he left it too late, killed most of the carrots along with the weeds and had to re-drill the crop. Since then he is more careful. The situation is complicated by the fact that burning off often conflicts with silage making, so that there is a temptation to do it too early or too late if Gerry Goodway, the contractor, arrives with his forage harvester.

A somewhat similar situation occurs with the potatoes which he planted last month. Again he is not allowed to use chemical weed control, so at least once a week during May he is down on his hands and knees scraping away the soil from the tops of the ridges to discover how long it will be before his potatoes come through. As with the carrots, leaving it until the last moment, he harrows the field before the potatoes come through to kill the maximum number of young weeds, then ridges the crop up again. He will go through the crop several times more to eliminate weeds before the potatoes meet across the rows in July and prevent further growth.

Control of weeds presents the organic farmer like Tony with a permanent challenge. It begins with endeavouring not to bring weeds on to the land in the first place by making sure that the manure has heated up enough to kill any seeds in it. The use of 'false' seed beds, as with the carrots, helps to exhaust the soil of its weed population – at least near the surface where they are most likely to germinate – and flame-weeding destroys yet more. This is followed by the tractor hoe which can be set to take out nearly all of the inter-row weeds, leaving only those growing among the plants to be removed manually. The final weapon in the armoury is the rotation, and here Tony is lucky in having a large proportion of his farm in grass. If the arable area becomes too weedy it can be ploughed and sown with grass and clover and another ley broken up for cropping. But it's a continuous battle; there are between 5000 and 10 000 weed seeds in the average square yard of soil.

Once silage making starts at Bridge Farm there isn't time for much else. Tony milks an hour earlier in the morning so that he can get through the chores before Gerry Goodway arrives, and he often doesn't begin the evening milking until 7 or 8 o'clock. But by working flat out they can usually finish the job in a week.

Choosing his time very carefully – and certainly avoiding silage making – the local NFU secretary usually pays Tony a visit during May. There's nothing special about this month; it's just that this is when Tony's farm policies fall due for renewal and the NFU secretary doubles as agent for the insurance society. Although he draws a commission of 10 per cent on the premiums he arranges, he knows that it's not in his long-term interest to persuade his clients to over-insure. The trouble is that Tony's expanding business, together with inflation, seem to have pushed up his bill for insurance in recent years.

Sitting round the kitchen table drinking coffee they go through the sections of his policy. As a tenant, Tony is not responsible for insuring the house and buildings – that's the landlord's job – but he would be rash not to insure all his livestock, machinery, feeding stuffs and other deadstock against fire. He must, by law, be covered for liability against claims from employees due to his negligence. He's also insured against liability towards the public – including a variety of risks ranging from damage caused by straying livestock to someone finding a piece of glass in one of Pat's yoghurts. So far, Tony has to agree, there's no room for saving on premiums. But that's the trouble with insurance, it all seems advisable.

'What about this "livestock in transit" – do I really need that?' asks Tony.

'Are you still taking calves to market?' Tony nods. 'And barren cows?' Tony nods again. 'You really ought to have it – it covers you for loading and unloading *and* at the sale yard. Easy for a cow to slip and break a leg.'

'"Theft of farming stock" – what about that?' demands Tony, anxious to find a saving.

'Well, if it was only the land here, I wouldn't recommend it necessarily. You're tucked away nicely here,' explains Mr Williams. 'But you've got that 10 acres up in the village with a bunch of bullocks grazing there – I saw them this morning on my way here. It wouldn't be too difficult for someone to back a lorry up to the gate one dark night. They could be up in Cumbria or down in Devon before you were awake. And then where would you be?' The logic seems inescapable and Tony has to agree that it would be imprudent not to insure.

Still looking for something he feels he could manage without, he settles on 'uncollected milk'. He's covered against his milk having to stay on the farm for any reason – perhaps because the tanker couldn't get through snow, or because he had inadvertently allowed milk containing antibiotics from a cow with mastitis into the bulk tank and the dairy might reject it. Is this the one to cancel? After all, he argues, he has always managed to get his milk away somehow, and he does have an emergency tank. And he could always turn it into yoghurt. And now he's organic, he doesn't use antibiotics very often. In the end he manages to convince himself that he can manage without this cover.

Mr Williams makes a note and then carries on quietly. '"Farm goods in transit". Now you're sending away all this marvellous stuff – vegetables and dairy produce – don't you think you need this?'

'But I'm already covered,' Tony protests. 'We've just been talking about it – markets and that.'

Mr Williams shakes his head and smiles sympathetically. 'Ah, no. That was "livestock" in transit. This is "farm goods" in transit.' Tony has to agree that they ought to be covered. Eventually they come to the end and he finds himself agreeing to an even bigger bill than last year. Mr Williams gathers up his papers and accepts an invitation to have a look at Tony's show beasts.

June

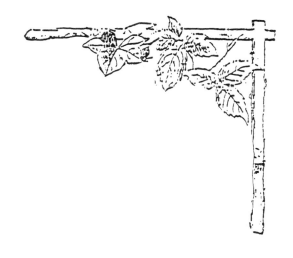

It's amazing how cold and dismal June can be at times in spite of its reputation as the hot and sunny month. It can be almost guaranteed to 'flame' at some stage, but by no means throughout its 30 days. However, the countryside is at its best this month, full of sound, scent and colour. Perhaps it's the June scents which are the most unforgettable, ranging from the delicate fragrance of the dog rose through the rich scent of the hedgerow honeysuckle to the overpowering, almost sickly smell of the elderflower. The most characteristic smell is that of new mown grass. Not, as Tom Forrest was explaining to some holidaymakers in The Bull who were waxing lyrical about it, new mown hay. 'You can't have new mown hay – it's only hay once it's cured,' he told them. 'Then it has an altogether different smell. It's grass, when it's just cut, as you've been enjoying.' The clackety-clack of the mower blade as it chats its way through the crop may be missing now that the grass is all cut by rotary mowers, but the scent is exactly the same.

That remarkable tree the elder adorns the hedgerows in June with its masses of creamy blossom and again in the autumn when laden with heavy bunches of purple berries. Prue Forrest's elderflower wine has been compared favourably to Frontignac, and her elderberry syrup mixed with hot water has eased many of Tom's colds. Non-alcoholic elderflower champagne delights the Ambridge children as it fizzes and bubbles in the glass while elderberry port, fortified with raisins, was one of Dan Archer's favourite winter tipples. As if this bounty were not enough, children still burn the pith from the centre of elder twigs with hot skewers to make pea shooters, and some of the older folk will hang a bunch of dried elderflowers in the house to repel insects. But no true countryman will use elder as firewood as it's supposed to bring bad luck. Bert Fry summed it up in one of his earthy rhyming couplets thus:

Elder burned in the grate

Means trouble soon or late.

Thunderstorms are quite common at this time of year. Farmers don't like them because they flatten the corn, and gamekeepers fear them

By June most of the pests and diseases of cereals have done their worst and the crops are in ear. But thunderstorms or drenching rain can still cause crops to be laid with devastating effect on yields.

OVERLEAF Flower-rich hay meadows such as this are rare. More than 95% have been lost over the last 40 years.

TOP Multiple suckling. Each cow has had two Limousin cross calves fostered in addition to its own.

ABOVE Stubbles after harvest. Most of these fields will be ploughed and re-drilled with cereals before winter sets in.

OPPOSITE Straw burning, a practice which has made farmers unpopular. It is now much more closely controlled.

ABOVE and TOP Oilseed rape, the crop which has changed the face of the countryside in spring. The acreage has increased 100-fold in 20 years. The tiny black seeds are crushed to produce oil.

OPPOSITE Modern machinery has taken much of the backache out of the harvest. Here straw bales are stacked in eights, ready for carting, to allow cultivations to proceed.

TOP Pools are now being created on many farms to encourage wild life of all kinds – and sometimes provide fishing.

ABOVE Red deer at grass. Deer farming is still in its infancy in the UK with fewer than 20 000 hinds (compared with about 18 million breeding ewes and 1.3 million beef cows).

TOP Flame weeding. Unable to use chemical sprays the organic farmer burns off the weed growth just before the crop seedlings emerge. Easy to mis-time and damage the crop.

ABOVE The machinery lines at the Royal Show in Warwickshire, mecca for farmers from all over the UK and abroad. The first show was held near Oxford in 1839.

OVERLEAF Bluebells thriving in coppiced woods. Carefully planned new farm woodlands should benefit both flora and fauna.

because they can jeopardise next season's shooting. The tiny pheasant and partridge chicks are very vulnerable and are easily drowned in a heavy downpour, or chilled, to die later of pneumonia. Another source of loss of game birds – adults, chicks and eggs – is the mower, cutting grass for silage or hay. This is carried out with such speed and relative silence compared with the old days that the birds get little warning of approaching danger; the tractor driver is insulated in his cab and unlikely to spot a nesting bird. Pheasants, partridge, rabbits, hares, fox and badger cubs and even baby deer have fallen victim to the mower or forage harvester.

The kitchen gardens are repaying all the hard work that has gone into them and yielding salads of all sorts – lettuce, radishes, spring onions – the first peas and early potatoes and, towards the end of the month, that quintessential symbol of midsummer, strawberries.

BROOKFIELD

Depending on the season, silage making can extend into June, leaving only a brief respite before haymaking begins. Hay is simply grass dried in the sun so that it can be stored, and until comparatively recently it was the only way for most farmers to conserve the summer's glut of grass for winter use. Gradually, since the last war, silage has taken over as the main method of grass conservation and some farmers now make no hay at all. The majority, however, are like Phil and while relying mainly on silage still make some hay. Among the disadvantages of hay are that it is heavily reliant on the weather, and no one has yet come up with a method of self-feeding it to stock, as they have with silage. On the other hand it must be much easier for Phil to take a few bales to outlying stock rather than a load of silage, and if he has a surplus a load of hay is much more saleable than silage.

Until recently many smaller farmers, those with limited livestock enterprises, were put off silage by the need to have a clamp and equipment to handle it. However, during the eighties a technique has been evolved which has brought many more farmers into silage making; this is making baled silage, sealed in a polythene bag. The big balers used by larger farmers for hay and straw are used to make the bales, which are then either sealed in a heavy gauge polythene bag or wrapped in the field in polythene bands to form an airtight package. Many a farmer, like Joe Grundy who never thought that he would come to rely on silage, now stores his winter feed in rows of shiny black 'sausages' in his yard. These are easily handled by tractor.

Even if they don't make as much hay at Brookfield as they used to, they still want to make it as well as possible. Good hay is an excellent feed, whereas poor hay is no better than straw, and a cow would have trouble eating enough of it to keep her in good condition. Hay has to be made reasonably quickly to retain its quality – ideally in 3 or 4 days. The longer it takes to make the poorer will be the feeding quality, especially if it rains which leads to moulds and the leaching of nutrients. Making less hay now than formerly, Phil and David can often avoid rainy weather, but it needs more than the absence of rain to cure; it also

Haymaking is not as important as it was before silage took over as the main form of grass conservation. The gangs of fifty years ago, turning by hand, have been replaced by one man on a tractor. But the crop is still at the mercy of the weather.

needs the sun. In dull, cool, overcast conditions without any rain, hay can take a week or more to make and all the time it is losing feeding value.

When Phil was David's age they tended to cut the grass and leave it until the top was more or less made before they turned it. Now, with better machinery and know-how, they turn it within hours of cutting and keep on shifting it until it's fit to bale. In perfect haymaking weather – sunny with a warm breeze – they have actually baled it in 48 hours, but that was exceptional. The contrast between haymaking at Brookfield now and when Phil was a boy is prodigious, as sometimes he can't resist telling David. In those days there would probably be eight or ten people, including children, involved in turning the hay and putting it into mows if the weather looked threatening, then pitching it on to horse-drawn wagons from where it was pitched off on to the rick. Neighbours like the Allards and Barratts would help and Phil's father would, in turn, help them when they had a field ready. So there was a large and happy if somewhat tired and sweaty gang to welcome Doris Archer when she arrived with the tea. Nowadays haymaking is not quite but almost a one-man job. David might mow the field, then turn it, then turn it again, then put it into rows, then bale it – all without getting off his tractor seat except to change the implement from time to time. The only occasion when the feeling of a team at work might be had would be towards the end with David giving the hay its final handling, turning two rows into one, Phil baling behind him and Bert starting to pick up the bales. But they are three men isolated in tractor cabs with their individual thoughts and observations.

'Everything all right, Neil?' Phil is making one of his regular visits to the pig unit at Hollowtree.

'Yeah,' says Neil unconvincingly, 'more or less.'

'What's the trouble?'

'It's this hot, muggy weather. Unsettles 'em. Come and have a look at this.' Neil takes him to a pen where a pig is lying stretched out, looking very much the worse for wear. Half an ear and the end of its tail have been torn off, and it has been savaged all down one side.

'How did this happen?' asks Phil.

'It's that pen up the end, Mr Archer. There's one beggar there will keep fighting. 'Course, once he sets in all the others join him. I had to take this one out or they'd have killed him between them.'

'They have made a mess of him, haven't they?' says Phil, peering in the pen again. 'It's something to do with the humidity, I reckon.'

'Thing is, it seems to be catching. There's another pen now where I got trouble.'

'Toys. That's what they need, Neil. Toys,' asserts Phil.

'Well, they got their chains,' says Neil, referring to the lengths of old chain from the muck spreader which he has fixed to the floor of the fattening houses to give the pigs something to relieve their boredom.

'Try some straw. Or blocks of wood. Something they can push around. That should take their minds off fighting.'

'Yes, Mr Archer,' says Neil doubtfully.

'And you must try and get this one back with the others as soon as you can. Otherwise they won't accept him, as you know.' Phil is referring to the fact that it's not possible to introduce a strange pig to a group, even if they are the same size, without trouble.

'Yeah, well . . .' is all Neil can manage.

'Don't suppose you have this trouble up at Willow Farm,' says Phil brightly. 'Keeping them outside.'

'Well, the sows and litters is all right. But when I brings 'em in to the fattening pens it's no different,' Neil explains. 'Pigs is pigs, when all's said and done.'

An essential though unpopular job towards the end of June is dipping. No one enjoys it, least of all the sheep, but it has to be done; it's the law. Compulsory dipping to prevent the spread of sheep scab began in this country in 1905. By the early fifties scab had been eradicated, only to be re-introduced 20 years later, it is thought through an importation of sheep from Ireland.

The long narrow dip is part of the sheep handling system which Phil installed at Marney's when he expanded the flock to 300. The sheep are supposed to slide into it automatically, but unfortunately some of them, remembering the last time, refuse to take the plunge and have to be 'assisted'. Once they're in the dip someone pushes each one under with a bar on the end of a long pole as they swim to the other end. They are supposed to be in the dip for a full minute and Jethro used to irritate

Phil and David by insisting on timing them. Eventually they climb up steps at the far end and enter a pen where they stay while the excess dip drains out of the fleeces and flows back into the tank.

During a tea break, Phil overheard Bert explaining the procedure to some holidaymakers who had stopped to see what was going on.

'We have to dip 'em twice a year on account of sheep scab,' he explained. 'Now, and again in the autumn. Mr Archer has to notify the authorities 3 days in advance, and then if they wants to they can send someone out to make sure we'm doing it right. He hasn't turned up yet

Farmers have been subjected to compulsory dipping to control sheep scab for most of this century. But with or without compulsion dipping is still needed to kill parasites.

but he might come any time just to check. Sometimes he'll take a sample of the dip to see if it's the proper strength. If it isn't, we have to dip 'em all over again. 'Course, it used to be the village bobby come to make sure we done it proper. I can see 'im now in the sun, sweating like a pig in 'is blue uniform, buttoned up to the neck. And 'is helmet.

'Sheep scab's a terrible thing. Caused by a tiny mite. Makes the sheep itch like fury; all their wool drops off and they gets terrible scabs. Someone in Ambridge had it last autumn. We all had to dip our sheep again. That was three times last year. Stupid thing is if *every* farmer in the country dipped *every* sheep properly just once – there wouldn't be no more sheep scab. Mind you, we'd still dip. It kills all the other bugs on the sheep's skin – lice, keds, ticks and that. And it kills the maggots, or puts the blowfly off, I don't know which. Maggots is just as bad as scab if they'm bad. They can eat a sheep alive. When we'm finished we chucks the dogs in an' all. Kills all their bugs too. Over where I come from someone threw the farm cat in as well, but that was a terrible mistake. You see a dog shakes hisself and runs off, but a cat licks itself. This dip is very poisonous and the poor old moggie died.

'Well, I'll have to get back. We still got half of them to do. There's over 700, including the lambs. Oh, and if that inspector chap don't turn up they can still get you. They takes a sample of the fleece in the market. If there ain't enough dip on it you have to do them again. Don't want that, do we?'

Phil felt that he couldn't have put it better himself.

By the end of the day everyone is covered in grease from the wool, sheep muck and dip. A tangy whiff of disinfectant fills the air. But everyone is secretly pleased that the job – unwelcome as it was – has been done.

HOME FARM

Brian may also have the tail end of his silage to finish – although silage making seems to be getting earlier every year – and later some haymaking and sheep dipping to contend with. But his main concern during the month is keeping an eye on his deer as they calve. One of his hinds is in trouble and looks as if she may need help. He has had her under observation for several hours and he and Sammy have tried unsuccessfully to get her into a building. There's only one thing to do now and that is to tranquillise her with a drug-filled dart. 'Darting' is something he contemplates only as a last resort as the drug tranquillises not only the hind but the unborn calf as well. He has been able to get near enough to see that the head and one foot of the calf have been protruding for some time. With any luck it just means that the calf has one shoulder back which is making it impossible for the hind to manage on her own; if they can handle her it should be easy to deliver the calf.

Jenny is just off to Borchester when Brian announces that he'll need her to come with him. The drug is quite powerful and if, by accident, Brian should inject some of it into himself while handling the dart – by no means an impossible thing to do – somebody has to be on hand to administer the antidote immediately. Jenny knows that there's no point in arguing. The hind and her calf are more important than coffee in Underwoods and, anyway, she's quite flattered that Brian relies on her for an important matter like this.

Having assembled his dart with the minimum amount of the tranquilliser, Brian picks up his mobile telephone – 'just in case' as he puts it – and he and Jenny go back to the deer paddock where Sammy is keeping an eye on the hind. Brian manages to get to within about 25 feet of the hind, and fires the dart into her haunch. She shoots off as if stung. After about 3 minutes the hind begins to stagger and then collapses. Sammy throws a sack over her head to calm her as she's still conscious. Brian quickly smears his hand and arm with antiseptic jelly, gently pushes the calf's head back into the womb and searches for the missing foreleg. Having located it he manages to pull out both legs and then very gently draws the calf out. Losing no time he injects the calf with the reviver and having detected some reassuring signs of life in it turns his attention to the hind. She is also given an injection of the reviving agent, and both are loaded into the stock trailer and taken back to the farm.

There are some nail-biting moments ahead as they know that the hind may, on coming to, decide to turn on the calf and kill it. It takes 3 or 4 minutes for the reviver to work. When Brian and Jenny (who has become too interested in the outcome to want to leave) peer over the back of the trailer the hind has recovered. The appearance of human faces seems to bring out the protective instinct in her and she moves to the calf. A moment later she is licking it and they know it will be all right. 'Thank goodness for that,' Brian remarks, looking at his wife mischievously. 'I thought for a moment that you were going to have another Bambi to look after.'

Two of Brian's chaps are busy re-straining a barbed wire fence which crosses a public footpath. This involves making sure that the wire is shielded, and Brian is checking that the job is being done properly. He knows that it's a path used by Lynda Snell and he doesn't fancy another bill for repairing a torn jacket. Like many farmers and landowners Brian often seethes quietly at the expense and inconvenience of maintaining rights of way across his land. He read recently that there were 120 000 miles of footpaths and bridleways in England and Wales, and more than 4000 miles in Borsetshire. Sometimes he feels, as he tangles with the

The tall fencing necessary to contain them have given red deer a somewhat forbidding aspect. Farmers who keep deer want it to be seen as a 'normal' enterprise.

indefatigable Mrs Snell, that a disproportionate number of them run across Home Farm.

Brian – and Phil Archer too, if they were really honest – regard rights of way across their land as an unwelcome intrusion, not so much on their privacy as their businesses. They also resent the many responsibilities placed on them for the maintenance of stiles and gates and the limitations imposed on their farming. The public on the other hand, personified by Mrs Snell, see footpaths and bridleways as their key to the countryside, a passport to be defended to the hilt. Farmers tend to argue that the network of rights of way is outdated and that many of the footpaths no longer 'go anywhere'. Those who use them believe that this alone adds to their charm and opens up huge stretches of country which would otherwise remain firmly closed. If Brian ploughs a field with a footpath across it, he is obliged by law to reinstate it within a fortnight. If he plants a crop it should not be sown where the path runs, and if it is drilled it should subsequently be sprayed with a chemical to kill it off. If the path or bridleway runs along the headland it ought not to be ploughed at all. Fines of up to £400 can be imposed for infringing the law on rights of way. Normally, however, persuasion is used first by the local authority, and only persistent offenders suffer the full force of the law.

If Phil Archer wanted to run a Friesian bull with his heifers he would have to avoid any field with a right of way running through it. Dairy bulls of any breed are not allowed, but the more placid beef bulls are permitted as long as they are running with cows or heifers. This suits Phil because, like most dairy farmers, he runs a beef bull with his heifers and relies on artificial insemination for the cows.

Some of the footpaths across Brian's land are virtually never walked and had more or less been forgotten, until the persistent Mrs Antrobus came round wielding her Countryside Commission do-it-yourself survey form. If the Commission succeeds in its aim of bringing the whole network back into use by the year 2000, Brian will presumably have no alternative but to reinstate them, although they may well subsequently fall into disuse. Phil tends to be tolerant of walkers while Brian regards them all as unwelcome intruders. But even he had to laugh in retrospect at the ill-considered accusation of a rambler with whom he had a difference of opinion as to the state of a footpath last winter. 'You can tell farmers don't like us,' the walker shouted, map in hand, as he proceeded on his way across a field of winter wheat. 'They always put their gateways in the muddiest part of the field.'

BRIDGE FARM

A busy time for Pat and Tony. Once the silage is safely in the clamp there are cabbages and leeks to be planted and the swedes to be drilled. It's a period of long days, rushed meals, late milkings, neglected children and the occasional bout of bad temper as they rush to get the jobs done. A 'deadline' hangs over them in the shape of another farm walk which, in a weak moment some time ago, they agreed to lay on later in the month.

The second week of June Tony fetches the bundles of cabbage plants from an organic grower in the next county and they set to work using an old two-row planter, bought from a friend. It's mounted on the back of the tractor and requires two people to sit on it and feed the plants one at a time into rubber-covered fingers on a wheel, which transfers them to the soil. It's a slow job – the tractor has to be driven in bottom gear – and takes several days. The first time they used the machine Pat and Art fed the plants in the wrong way round, with the result that they had two rows of impeccably planted cabbages with their roots waving in the air. Fortunately no real harm was done and they all had a laugh at their efforts before collecting the plants up and having another go.

As soon as they have finished the cabbages they start on the first batch of leeks; they plant some this month and some in July to lengthen the harvesting period. Tony notices that there are quite a lot of thistles coming in one corner of the spring wheat and organises a small gang of women to pull them by hand.

He is fortunate in being the only farmer in the area who relies on unskilled casual workers. With the rise in living standards and fewer working families in the villages these days, farmers and growers are finding it harder to recruit labour. Fortunately, on conventional farms modern techniques of precision drilling and selective herbicides have done away with much of the handwork, leaving workers available for the organic farms. No one needs to hand-hoe Brian Aldridge's sugar

Planting out cabbages. In spite of modern technology, vegetable growing remains a labour intensive business. On an organic farm, controlling weeds without chemicals intensifies the situation.

beet but Tony still needs casual labour, not only for weeding but for harvesting and packing his crops from time to time. He can usually find enough women willing to come between 10 o'clock and 3.30 pm, which is when most of his field work is carried out. They often need collecting and taking back, together with their prams and young children, and they have to be constantly supervised. Most of the jobs have to be done on an hourly rate rather than piece work so they need plenty of encouragement to keep at it.

At last the day of the farm walk arrives. Tony and Pat have finished off several weeks of feverish activity by tidying up the yard and buildings, pressure hosing the tractor, which gleams like new, and mowing the lawn. With dire memories of the last time they held an event of this kind, when Grey Gables laid on an 'organic' lunch which totally failed to satisfy the visitors' appetites, Pat and Tony decide on a 2 pm start, although the invitation in the regional organic news-sheet ended, 'Why not bring a picnic and eat it in the garden?'

A crowd of about 30 has turned up and is wandering round the yard, peering over stable doors pointing things out to one another. They don't much resemble the cartoonist's impression of a group of organic enthusiasts, but on the other hand they look altogether too earnest for a bunch of conventional farmers. There are one or two beards among them, unusual in farming circles, and a couple of dedicated older women with white hair scraped back in tight buns – but no sandals. It has been raining and looks like doing so again, so green wellies and well-worn Barbours are the order of the day for most.

Tony climbs on to a trailer and bids them all welcome. 'It's not the best time to be showing people round the farm,' he shouts, unnecessarily loudly. It never is, of course; no farmer would admit that. 'There are still plenty of weeds about, and if you come across any I'd be glad if you pulled them up.' This draws a laugh which allows him to be more serious. 'We started converting to the organic system five years ago, bringing in 30 acres a year. Last year we brought in the final 50 acres in one go. The mainstay of the farm is still the dairy herd – 65 Friesians – but we grow a wide range of crops all carrying the Soil Association symbol including spring wheat, carrots, swedes, leeks and cabbages. Pat's got some leaflets with the details on for anyone who'd like one. The farm is . . .'

'Why did you go organic?' Tony's introduction is interrupted by a question from a thin-faced girl with a pointed nose and glasses. Pat later identifies her as a member of the Women's Group from Borchester Tech. Tony flounders.

'Why did we go organic?' he repeats the question slowly. 'Now that's a good question.'

At that moment Pat speaks up from the edge of the crowd. 'I think I can answer that.' Everyone turns towards her, including a relieved

Tony. 'We were worried about what was happening in agriculture generally, as well as our own farm. We were milking 100 cows and talking of going up to 120 before quotas came. We were plastering the grass with nitrogen and polluting the streams, we were up to our knees in slurry, the vet seemed to be here half the time – all to produce milk of doubtful quality which nobody wanted. We decided to reduce the herd, not to push the cows so hard, cut the nitrogen and start growing the sort of crops people were looking for. It just seemed the right thing to do.' There are murmurs of approval all round except from the girl with glasses who enquires, 'But philosophically?'

Tony looks up at the sky. 'I think we'd better move off before we get any more rain. It doesn't look too promising,' and he jumps down from the trailer. At that moment he spots Eddie Grundy coming into the yard. 'What on earth does he want?' he asks Pat.

'Cause trouble, I expect,' comes the reply. 'You know Eddie.'

'Hallo, mate,' says Eddie as he approaches. 'All right if I come along? I'm interested in this muck and mystery.'

The group sets off round the farm with Tony in the lead, stopping at each crop.

'These carrots were drilled on 20 May and we flame-weeded them on 30 May. We shall hand-weed them in about 3 weeks.'

'How do you keep the carrot fly off?' someone asks.

'We just find that drilling around that date we don't seem to be troubled,' Tony replies. 'Right, let's look at the spuds.'

After the crops have been examined the party moves on to look at the grass fields, where some members get down on all fours looking for traces of the herbs which Tony assures them were included in the seed mixtures. And finally they reach the cows themselves, munching contentedly on a clover-rich ley. Tony gives details of stocking rates, milk yields, butterfat content and feeding regimes, pointing out how important it is to obtain some sort of premium for the organic milk if it is to compete with conventional dairy farming.

'What do you do about summer mastitis?' asks a visitor, referring to one of the worst scourges of the dairy herd, a deadly disease spread among dry cows by flies. Conventional farmers simply squirt a long-acting antibiotic up each teat after drying off, but organic farmers are not allowed to do this.

'I get 'em in twice a week and smear Stockholm tar on their teats. I don't know whether it keeps the flies off the cows but it certainly keeps 'em off me!' Tony jokes.

The visit finishes with tea back at the house followed by a vote of thanks given by one of the elderly women, who praises Pat and Tony for their courage, acknowledges how much hard work must be involved and thanks them for allowing them to see their farm 'warts and all'.

July

By now, the countryside has begun to take on a full-blown, almost blowsy look. The grass is not so green, the leaves on the trees look tired, and although there are plenty of wild flowers to adorn the hedgerows the cherry, the may, and the elder have blossomed and are now setting their seed. The birds are less loquacious and, having reared at least one and probably two broods, start to moult; some of them, including the cuckoo and swift, even think of leaving Ambridge for their winter quarters in Africa.

It's the season of those hardy annuals of high summer – the Royal Show and the Game Fair, Henley and Wimbledon, Cowes Week and Goodwood, and nearer home, the Ambridge Fête with its yearly squabbles over who is to run the white elephant stall, whether to have a fortune teller and how to prevent Mrs Snell from hogging the tannoy. School holidays begin and the more fortunate youngsters get ready for Pony Club camp. It's a time for taking a picnic down to the Am, sitting at the water's edge surrounded by purple loosestrife and willow herb, watching the dragonflies and hoping for a sight of the elusive kingfisher.

For farmers July sees the beginning of the cereal harvest, though it was not always so. That keen observer of the agricultural scene Thomas Tusser makes no mention of cutting the corn in his advice for this month.

Go, sirs, and away to ted and make hay,
If stormes drawes nie, then cock apace crie.
Let hay still bide till well it be dride.
Hay made away carrie, no longer then tarrie.

July storms obviously plagued the farmer just as much in the sixteenth century as in the twentieth, and there are few more disheartening sights than a promising field of wheat or barley laid flat.

In spite of the total protection afforded them by the Wildlife and Countryside Act 1981, bats flourish in far fewer numbers than in Phil Archer's youth, although they are still to be seen flitting round the yard and garden at Brookfield. They are feeding their young at this time of year and consequently on the wing for long periods gathering insects. Walter Gabriel's tired old eyes were unable to pick them out the last time he came to supper – not long before he died – but that didn't stop him recalling a trick which he, Sam Blower and Silas Winter had played on the local WI in the twenties.

'They was performing a Chinese play in the village hall,' he wheezed. 'Hardly any speaking – just nodding and bowing to each other, all dressed up in them long silken robes. Terrible serious it all was – until we got there.' Walter broke into a fit of cackling and it was some time before he could continue with the tale.

'We caught three bats – picked 'em up off the roosts – and took 'em down the hall and loosed 'em through the fanlight over the door. Oh dear, oh dear. You never saw such bewilderment in your life. First the women in the audience goin' mad as the bats swooped over their heads. Then the actors as they reached the stage. They had to turn round there and come back – that meant more trouble. Everyone clasping their heads and screaming. Then back up the stage again. All them folk in funny clothes – they did some bowing then, I can tell you,' and he collapsed into another burst of uncontrolled mirth. 'We tried to get your Dad to come along with us,' he went on when he had recovered, 'but he said they had a cow calving. Doh, I never laughed so much before or since. They never found out who it was. There was them as had their suspicions but nothing was ever proved.'

It's a sobering thought that if anyone did that today and was apprehended they could be fined £3000 – £1000 for each bat.

BROOKFIELD

From the beginning of July, sometimes even earlier if the weather is muggy, Phil listens carefully to the radio each day for a 'blight warning'. Potato blight destroys the foliage prematurely, leading to small tubers which sometimes also carry the infection and won't store. Once conditions are favourable for the spread of the disease Phil sprays his potatoes at fortnightly intervals with a protective fungicide.

July usually sees a flurry of activity at Brookfield as the silage clamp is uncovered to receive a second cut of grass. The fields from which the first cut was taken at the end of May will have been fertilised, but the season will dictate when the second crop is ready to cut. This time the job is accomplished within a week.

Although Phil will have made most of his hay in June there's always

Potato blight, which caused the great Irish famine in 1846, is still an ever-present threat. Most farmers spray their crops regularly in late summer to prevent it.

one patch he leaves until July. This is part of an old 'hay meadow' he bought when Meadow Farm was split up in 1979 and it's very rich in wild flowers. Such meadows are becoming rare, and local conservationists pleaded with him to leave it as it was, not to plough it up or apply any fertiliser. Further pressure was applied by the local Farming and Wildlife Advisory Group who, when Phil pointed out that he had just paid £1700 an acre for the field and needed to get a good return in order to help pay off the mortgage, suggested that he left 3 acres at one end of the field in its original state. They were joined by villagers, including his own father Dan Archer, who remembered walking the footpath which ran across the field and admiring the cowslips, harebells and knapweed, the oxe-eye daisy and the spotted orchids, the vetches, the bird's foot trefoil, ladies' smock and cranesbill as well as the butterflies and grasshoppers. Fed up with all the fuss Phil announced that he would leave the whole field as it was but after a year or two, as he thumbed through his accounts, and goaded on by David, he decided to follow the FWAG advice and leave the 3 acres next to the footpath undisturbed. The rest of the field responded to heavy dressings of fertiliser, produced big crops of hay and provided plenty of aftermath grazing, but the flowers disappeared. Now, as a gesture of goodwill to the conservation movement, he puts off making hay on the small patch until well into July – after most of the wild flowers have bloomed and seeded. By this time, of course, the grass has all headed up and Phil harvests a poor crop of hay, fit only for bullocks.

Now, ten years after acquiring the meadow, David still nags at his father. 'How much longer are you going to starve those 3 acres of fertiliser, Dad? That hay's only fit for bedding, anyway, it's so mature. Have you worked out how much it costs us per bale?' Phil did work it out once and didn't much like the result. He was on the point of deciding to give up preserving it as an old-fashioned hay meadow, at one stage, when the Borsetshire Nature Trust approached him with a tentative offer to buy the land so that they could 'manage' it. He had totally unmerited visions of bands of enthusiasts trying to cut it with scythes while hordes of conservationists scrambled round on their hands and knees examining the flora and fauna – and decided not to sell. The trouble was that having turned down their offer he found it exceedingly difficult not to carry on managing the meadow in the old-fashioned manner himself. After all, he argues, it is only 3 acres out of 460. Occasionally, he and Jill take a walk there to admire the orchids and agree that they should continue to preserve it – 'for the time being, at least'.

After more than 20 years of ripping out hedgerows, felling trees and filling in ponds, many farmers are now becoming keen conservationists. Even the Ministry grant structure has been reversed, and instead of being paid to remove hedges farmers can get financial help to establish new ones. In the forefront of the changing attitudes is the Farming and Wildlife Advisory Group which now has a branch in nearly every county. Farmers, landowners and naturalists have got together specifically to show farmers how they can conserve wildlife and the landscape without damaging their farming interests.

A visitor to Brookfield at this time of year may be surprised to see some of the cows with red tapes round their tails. These are cows which are being 'dried off' to give them a rest before they calve again in the autumn. Phil had a bunch of school children round recently and was trying to explain his dairy policy in simple terms. When he began describing how they dried cows off in the summer a freckle-faced ten-year-old exclaimed, 'But mister, I thought you *wanted* 'em to give milk.'

'We do . . .' Phil began patiently, but he was interrupted.

'Well, why d'you try and stop 'em, then?' demanded the youngster, logically.

Phil went back to square one. A cow only gives milk when she has a calf, he explained. The farmer takes the calf away after a few days so that he can sell the milk and rear the calf on a cheaper milk substitute. For the first 4 to 6 weeks her milk yield goes up and up; then it levels out and begins to decline slowly. If he keeps on milking her, he told the children, she would carry on giving milk for quite a long time – another year or more. But by the summer she would only be giving perhaps 5 or 6 litres a day, hardly enough to justify putting her through the parlour. Meanwhile, he has arranged for her to have another calf in September when with luck she would yield 25 litres a day. It's important, he added, to dry her off to give her a break of a couple of months before she calves again.

The Brookfield corn harvest usually starts around 20 July. Two factors influence the date. One is whether the barley is ripe, and the other is whether it is dry enough to combine. It's frustratingly feasible for Phil to have a field of corn ready to cut but the weather making it impossible to start – or perfect combining weather but nothing ripe enough to harvest. For the week before the anticipated day they are busy with preparations: adjusting and greasing the combine, tidying up the grain store and cleaning out the pit into which the corn is tipped as it comes from the field.

Once harvest gets under way it's by no means plain sailing. It can be held up by mechanical breakdowns or waiting for crops to ripen but more often by the weather. Twenty minutes' rain in the morning can destroy any hope of harvesting for the rest of the day. If the straw is too wet it won't go through the combine properly, and after prolonged rain the land may be too wet to take the machine. But the main reason for holding off is that the grain itself is too wet. Phil stores most of his corn and unless it's dry it will heat up and go mouldy. It's true that, like most cereal growers, he has a dryer – but drying wet corn is a very expensive pastime. He prefers to wait until the sun and wind have done the job for him and saved him an inflated electricity bill.

That's why, even in good weather, combining doesn't usually begin until 11 or 12 o'clock each day when the dew is off and the ears have had a chance to dry out. Phil positions himself at the dryer, grinds tiny samples of the corn and rams them into his moisture meter to see whether he needs to blow heated air through or whether he can get away with cold air or – on rare occasions – if it is dry enough to go straight into the store without treatment. They normally carry on combining, with a short break for lunch, until about 8 o'clock, unless it's one of those warm evenings with a good breeze when the corn is dry enough to carry on after dark. The combine driver would 'know' when to knock off; he'd feel that the straw was getting too damp for the machine to work efficiently.

As with haymaking, except for bale handling, there is virtually no manual effort involved in harvesting corn these days. The combine cuts the crop and threshes it, sending the corn into a tank and spewing the straw out behind. When the tank is full the driver simply presses a button to discharge it into a trailer, which is then taken back to the corn store. This is a far cry from the days of the binder and thresher when the crop was manhandled half a dozen times. The combine harvester is a comparative newcomer to British farms. Although there were a few at work in the twenties it was not until the sixties that they finally ousted the binder. The worst aspect of combining then was the dust, but even that has been largely overcome by the introduction of the modern cab.

In most seasons, the Brookfield harvest progresses in fits and starts as it bows to the fickleness of the English summer. In theory, with about 200 acres of combinable crops – wheat, barley, oats, rape and beans – Phil could get through the whole lot in a fortnight; in practice he is more likely to be at it for 6 weeks. Fortunately, once harvest has started there are plenty of jobs to be getting on with when conditions are not suitable for combining. As soon as the corn is cut the straw can be baled and carted and the field ploughed and cultivated ready for another crop.

HOME FARM

A firm date for Brian at the beginning of July is the Royal Show held at Stoneleigh in Warwickshire, only about 40 minutes' drive from Ambridge. This annual showpiece of the Royal Agricultural Society of England started 150 years ago, and until it came to its present site in the early sixties it travelled to a different venue each year. Stoneleigh has now become a centre for many agricultural organisations and breed societies, and the site of other more specialised farming events throughout the year; but the Royal retains its attraction as a social and farming event and attracts nearly a quarter of a million visitors (many of them from overseas) over its 4-day run.

The show is still quite a hierarchical occasion, and in spite of the fact that the entrance fee is now the same throughout the week (it used to start much dearer) the Aldridges of the farming world still tend to go on Monday or Tuesday and the Grundys on Wednesday or Thursday. Brian lunches in the Governors' Pavilion, Phil in the Members', while Eddie is more likely to be found with a pint of beer in The Cornish Elm, a 'pub' conveniently sited near the stock lines. Brian, like most enlightened employers, encourages his men to go and gives them a day off and a ticket so that they can take an interest in the latest technology.

Brian usually has one day at the Royal on his own and one with Jennifer. If he sometimes wishes that he could have two days to himself it's not because he doesn't enjoy his wife's company, it's simply that there's so much to see and he resents being dragged off to look at things which have beguiled her. They begin with a cup of coffee in the Governors' Pavilion while they decide what to do. Brian wants to go round the arable plots – he's particularly interested in seeing the new pea varieties – visit the farm woodland demonstrations and look at a new big combine. Jenny favours the flower show, the food hall and a look round the Jacob sheep. They arrange to meet for lunch and Jenny disappears to see to her hair while Brian strides off purposefully.

Brian gets back at the agreed time and is sitting thumbing through a brochure on rough-terrain forklift trucks when Jenny arrives carrying two bags.

'What on earth have you got there?' he asks.

Jenny puts her parcels down. 'Darling, you'll never believe, I've got some trout for your supper.'

'Trout? We've got trout in the freezer, haven't we?'

'Not like these. I just couldn't resist them. And I've got some cheese and some oak-smoked ham. It all looked too delicious.'

'Surely it'll all go off in this weather.' Brian's eyes narrow suspiciously. 'Anyway, what's in the other bag?'

Jenny fingers the carrier bag. 'You'll never guess. It's a pan for poached eggs. Ours is worn out.'

'But Jenny,' Brian sighs. 'You could have got one of those in Underwoods. Now you've got to carry it round all day.'

'Brian,' Jenny wheedles. 'I thought you could speak to that nice friend of yours, Jerry, and ask him to keep them on his stand for us.'

Brian knows when he's beaten. 'Oh, I suppose so. Anyway, let me get you a drink. Pimms?'

Over a lunch of smoked salmon, cold duck and strawberries and cream, Jenny demands to hear about everything that Brian has seen – until he actually begins to tell her.

'That new combine looks terrific,' he enthuses. 'I sat in the cab for ages playing with it.'

Jenny's looking around. 'Isn't that the woman from Shropshire, we met last year?'

'It's got press button control of everything and you can call up virtually any information on the TV screen – even tells you the weight of the crop you're cutting. Here, look at this.' Brian produces a slip of paper from his pocket. 'That's all the details of the 50-hour service – look, greasing, oiling – all you do is select what info you want, press a button and you get a printout. More wine?'

'Sounds amazing,' says Jenny, and then continues, 'Darling. Can we have a house cow?'

'A *house* cow?' Brian can hardly believe his ears.

Jenny carries on unperturbed. 'I was looking at the Jerseys. They've got the most adorable eyes. And the milk's so rich.'

'It's out of the question, Jenny.'

'But darling, think what it would save – and Kate and little Alice would love it.'

'And who's going to milk it?' Brian snaps back.

'Sammy could milk it, couldn't he? Isn't that what he's for?'

Sometimes the most effective way of tackling an outbreak of disease is to spray it from the air. In the past Brian and Phil have combined to reduce the cost of application.

Brian looks pityingly at his wife. 'Sammy's got his hands full with all the ewes and lambs, not to mention the cattle and the deer. Anyway, what about weekends?'

'But darling, they look so lovable. Come and see them. I know you'd fall for them . . .'

The combine harvester discharges its tankful of barley into the trailer. Experienced tractor drivers receive it on the move; the unpractised wait until the combine stops. Although cutting is still in progress some of the straw has already been baled.

The corn harvest at Home Farm begins, if anything, a little earlier than at Brookfield. This is because Brian has lighter land than Phil which encourages the winter barley to ripen quicker. It's also because once he makes a start there are fewer hold-ups since, with nearly 1200 acres of cereals, rape and peas, there is nearly always something ready to cut. With three combines Brian can also keep going even if one of them does have a seized bearing; at Brookfield this would bring harvesting to a total standstill.

Brian will still be keeping a close eye on his deer this month although most of them will have calved, and he'll be getting the first of the lambs from his second lambing off to the butcher. Apart from these jobs, and perhaps some spraying of the sugar beet to keep the aphids at bay, the earlier half of the month is relatively slack and a traditional time for a break. If Brian doesn't go away himself he will encourage his men, especially the younger ones without children, to take part of their holiday entitlement before harvesting.

In 1988 the EC introduced a system known as 'set-aside' in an attempt to reduce surplus production. The term was coined in the USA where, for many years, farmers have been paid by the government not to grow crops. In this country if a farmer decides to take advantage of the offer to fallow part of his holding the government will pay him up to £80 an acre. But he must agree to stop growing crops on at least 20 per cent of his established arable acreage for five years, with an opportunity to opt out after three.

To an outsider it must appear like money for old rope – being paid for doing nothing. Indeed, an American farmer is alleged to have been so enthralled with the idea that he wrote to his Congressman asking whether there was any chance of being paid for *not* raising hogs on the corn which he was *not* going to grow. But it is not as simple as it might appear. Some farmers have 'fixed costs' which actually amount to more per acre than the payment offered. These are costs such as rent, labour and machinery which are unlikely to vary much with a reduction in the arable acreage. It's because of their fixed costs that farmers are always on the lookout for more land over which to spread; most could cope with 20 per cent additional acreage with their existing labour and machinery. If they cut the acreage by a similar amount, they might use less fuel, fertiliser and seed but they would be unlikely to manage with fewer men or machines, unless the farm was very large. To qualify for the annual payment the land has to be kept in good condition as fallow or woodland or used for non-agricultural purposes. Farmers are not allowed to graze it as this, it was felt, would be unfair to other livestock producers. It is reckoned that it would probably cost half of the £80 an acre to plant it with a suitable grass mixture and keep it mown to control weeds – or to cultivate it regularly if it is not seeded.

Brian's initial reaction to the set-aside proposals was that the payment was too low to persuade him to give up growing crops since, even with falling prices, he could still make considerably more profit than £80 an acre – let alone the £40 an acre he would be left with after managing his fallow. He agreed with the general consensus that it was only likely to prove attractive in respect of poorer land incapable of producing good yields. For this reason he felt that set-aside would not be likely to result in a big cut in production, especially since anyone taking advantage of it was likely to concentrate their efforts on increasing production from the other 80 per cent of their land. However, along with some other farmers in the neighbourhood he registered with the Ministry of Agriculture – just in case he wanted to take advantage of the scheme in the future.

BRIDGE FARM

Since Tony and Pat added vegetable growing to their dairy enterprises they miss the traditional respite enjoyed by other farmers in between haymaking and harvesting. Their situation is aggravated by their constant war with weeds. Apart from their potatoes, which this month will be meeting across the rows and so preventing further weed growth, they have about 12 acres of crops requiring regular attention to keep them clean, and a further 2 acres of leeks to be planted during July.

After standing still for several weeks the cabbages planted at the beginning of June have begun to move, and Tony goes through them for the first time with the tractor hoe, setting the blades so that they take out the maximum growth of young weeds without damaging the plants. It's a job which requires hard concentration and any distraction can lead to errors. In spite of Tony's efforts to hide any damage, little

Ragwort growing in profusion. The handsome yellow-flowered plant is an official 'injurious weed' and poisonous to stock. Because the animals naturally avoid it, ragwort tends to spread unless destroyed.

escapes Pat's questing eye, and it's not long before he's challenged about 'that bare patch'.

In the second week of the month they bring in a gang to hand-weed the baby carrots. This time it's Tony's turn to get excited as he moves among the workers making sure that they don't pull up carrots along with the weeds. Although the young plants are only about 2 inches high, their bright orange roots are hard to conceal from him. The second planting of leeks entails setting up the irrigation pipes again, not one of Tony's favourite jobs. The first year he grew cabbages it was very dry and he nearly lost them. It convinced him of the necessity of having some form of irrigation to establish his crops, and the following autumn he bought a job lot of sprinkler equipment at a farm sale for £50. He draws the water from Heydon Brook, which flows through his land, using a tractor-powered pump. Coupling up the lightweight aluminium pipes is straightforward but operating the system subsequently can present difficulties. Tony has only enough equipment to water a quarter of an acre at a time, which means moving all the pipes every 6 hours. If the previous move was at, say, 6 o'clock in the evening this means a walk across the fields with a torch at midnight to switch the tractor off. This sort of thing can be planned in advance, unlike blockages which Tony feels always happen at the most inconvenient moments. The trouble is that when the irrigation equipment is lying idle, slugs and snails crawl inside the pipes. Sooner or later they get swept along to one of the sprinkler nozzles and block it. Tony ought to have learned by now that the odds against unblocking a nozzle without getting soaked are quite high. This is uncomfortable enough during working hours, but it's surprising how often it seems to happen just as he's togged up for a wedding or to go down to The Bull.

'Tony, there's a nozzle blocked on the irrigation. I've just been across there,' shouts Pat across the yard.

'Well, it won't hurt until I get back.'

'Of course it will, Tony, those leeks are getting no water at all,' Pat insists.

He should have changed before he went, or at least put on gumboots and an oilskin. His cavalry twills were plastered up to the knees before he even reached the offending nozzle. He managed to unblock it fairly easily, but when trying to screw it back it was suddenly torn from his hand by the force of the water and hurled into the air. By the time he had found it 20 feet away he had been thoroughly soaked by the adjoining sprinklers which were still working very effectively. On his way back he managed to trip over one of the pipes and measure his length on the ground. But despite the occasional soaking – and the fact that weeds like water too – Tony knows that irrigation is essential for his crops to get a good start.

'Daddy, Daddy, come quickly. There's a dead calf in the orchard.' Tommy rushes into the yard where his father is sitting on the steps of the milking parlour in the sun, changing the teat cup liners.

'Are you sure?' he asks. 'You're not teasing?'

The child tugs at his arm. 'No, Daddy. It's up the top by the pear tree.

I was looking for wild strawberries. Quick, come on.'

The under-sized calf is lying against the hedge showing no signs of life; the distraught cow moves off when they approach, then comes back almost pushing past them to get at the body.

'Oh, God,' says Tony to himself, 'looks as if she's aborted.'

Tommy peers up anxiously. 'What did you say, Daddy?'

'Looks as if she's had her calf too soon,' he explains. 'Come on, we must go and check up.'

A glance at the chart on the dairy wall confirms his suspicions. No. 77 is not due to calve until 21 September, another two months. There are several reasons why a cow might abort – she could have been fighting or eaten a poisonous weed or been fed on mouldy hay or, and this is what Tony is dreading, she could be infected with brucellosis. All abortions in cattle are supposed to be reported to the Ministry of Agriculture veterinary service and Tony accordingly goes to ring the local office. He is urged on in this direction by a strong desire to know what caused his cow to drop her calf prematurely. Before long his local vet, Martin Lambert, working on behalf of the Ministry, calls to take samples. Tony is 99 per cent certain that it isn't brucellosis, but he has to wait between 7 and 10 days for the result of the test; meanwhile, of course, he has lost a valuable calf.

Brucellosis is one of a number of major livestock diseases which have been all but eradicated from this country in Tony Archer's lifetime. In its day it caused terrible suffering to both cattle and human beings. It led to widespread abortions among cows and brought many farms to financial ruin. It also led to the spread of undulant fever, especially among farmers, cowmen and vets who had close contact with infected foetuses; the depressive effect of this disease drove a number of sufferers to commit suicide out of sheer desperation. After a 10-year eradication campaign orchestrated by the Ministry of Agriculture, brucellosis, or contagious abortion as it was often known, was more or less eradicated towards the end of the seventies. Bovine tuberculosis is another disease which took its toll of both man and beast. Although human beings were protected from catching it to some extent by the pasteurisation of milk, the disease continued to cause great loss of production in cattle until it was finally brought under control in 1960. It still lingers in some parts of the country, where badgers are accused of harbouring it. Because of the dangers to stock and public at large, cattle are continuously monitored for symptoms of both these diseases.

Some diseases are regarded as 'notifiable' – that is any farmer who suspects his stock may be infected is by law obliged to inform the Ministry of Agriculture, the local authority or the police. In addition to the two mentioned above they include foot and mouth disease, which ravaged the country so devastatingly during the winter of 1967 to 1968, swine vesicular disease which caused the slaughter of Phil's entire pig herd in 1974, and sheep scab, which brought poor Joe Grundy to the magistrate's court recently.

August

The corn harvest in Ambridge *can* start as early as the middle of July and *may* go on until the middle of September, but the quintessential harvest month is August. Except in totally aberrant seasons such as the drought year of 1976 the combines will be crawling through a crop somewhere in the neighbourhood throughout the month.

If the countryside is beginning to look a little tired the gardens and orchards are certainly making up for it with their harvest of vegetables and fruit. There are tomatoes and cucumbers, blackcurrants and gooseberries and marrows galore in Tom Forrest's garden and greenhouse, along with beans – runner, French and broad – young carrots and all the salads. Out in the orchards are the first of the windfall apples for Prue to cook with the earliest blackberries – and the wasps are already busy at the plums and greengages. Wasps are a sure-fire topic of conversation at The Bull at this time of year, how to find a nest and how to destroy it being the two main subjects for debate. Joe Grundy believes in catching a wasp and coating it with flour. 'He'll soon lead you to his nest,' he asserted the other night. But Tom believes in catching one and tying a bit of coloured wool round its waist. 'That way,' he explained, 'it slows him down. He still goes back to his nest but slow enough so you can follow him.' Joe laughed and said. 'We ain't all on our walking frames yet, Tom Forrest,' but Tom was in a genial mood and laughed it off.

The thrush is back in Ambridge after his silent moult, singing on the topmost branch of the tall cypress in Brookfield garden. In a spare moment Phil and Jill might sit in the evening listening to its sweet song as the swallows wheel overhead. The swifts are getting ready to migrate again – those incredible birds alleged to be able to eat, sleep, drink and mate on the wing – the young miraculously making their way to Africa without help from their parents.

BROOKFIELD

It's 6.45 am and Jill is about to get out of bed. She doesn't want to but she knows that at this time of the year it's advisable if she's to get everything done before it's time to go back to bed again. Phil got up at just after 6 and listened to the farming programme while he shaved. At 6.30 he took Jill up a cup of tea and went across to fetch the cows in for milking as Graham is on holiday.

By 7.15 Jill has got the first lot of yesterday's dirty washing in the machine and started on the housework. At 7.30 the telephone rings; it's Neil asking whether he'll be needed when he has finished at Hollowtree. Jill asks him to hang on while she goes over to the milking parlour to ask Phil, remembering to take the can to bring some milk back for breakfast. The answer is 'Yes please, he can carry on with the baling.'

Phil comes in for breakfast at 8.30 and begins opening the post as he strolls round the kitchen.

'Phil, don't leave your letters all over the kitchen, there's a dear. I've just tidied it up,' says Jill gently. At that moment the door opens and in comes David.

'Dad, can you come and have a look at this baler?' he asks.

'Morning, David,' says Jill. 'Have you had any breakfast?'

'Oh, morning Mum. Yes, I've had mine, thanks.'

Phil suddenly explodes. 'Twenty-three,' he shouts, waving a paper from the Milk Board. 'Look at this, David. The TBC's up to 23.' He looks at the form again. 'Must be a mistake. We haven't been as high as that for ages,' he adds, referring to the total bacterial count of the milk, regularly monitored by the MMB.

'Dad, the baler. I can't sort that knotter out. Can you come and have a dekko?'

'Yes, all right,' Phil replies and follows David out of the door.

'What about your breakfast?' wails Jill.

'Back in a moment,' comes the answer, and Jill sighs as she puts his plate in the bottom of the Aga.

By 9.30 she has cleared up the breakfast and is busy sorting out lunch when there's a tap at the kitchen door. It's a fresh-faced young man in a sports jacket and light twill trousers who announces that he's the new rep for Finch and Tanner, the dairy suppliers. Jill thinks quickly. David's wrestling with the baler, Ruth's washing down the parlour, Bert's greasing the combine and fuelling the tractors and Phil's off looking round the stock. She explains kindly that he has called at a rather busy time and suggests that he comes back after harvest. He's just disappearing out of the gate when Phil arrives back in the Land Rover.

Back at the grain store the corn is tipped. If it is too wet it will have to be dried before storage. Farmers try to avoid this expense by not combining in damp conditions.

'Who was that going off in the Escort?' he enquires.

'Oh, the new man from Finch and Thingummy, you know – the dairy people. I told him to come back after harvest.'

'Oh, blast. I wish he'd hung on. I desperately wanted to see him. We're right out of teat dip.' Phil stalks off across the yard towards the grain store. Half an hour later he's back. 'Jill, Jill, where are you?'

'Out here,' she shouts, 'digging some potatoes.'

'Look, we're going to make an early start. The dew's lifted.'

'I expect you'd all like some coffee. I'm just coming.'

Ten minutes later they set off for the last of the winter barley in Upper Parks, but not before Phil has left a list of instructions. If the corn merchant rings she's to say there won't be any more barley but they'll be into the winter wheat in about a week. There's a chap bringing some diesel – she's to have it put in the far tank – and there's a cow looks like calving, down in The Grove with the other dry cows. It's No. 483, the one with the white head. Can she slip down and have a look at it before lunch?

'Where will you be, Phil? If I need you.'

'Oh, not far away – either up at the grain store or out in the field. Depends,' is the enigmatic reply.

Jill's putting the finishing touches to a gooseberry tart when the telephone goes. It's the local radio station. They want to ride round in the cab of the combine and interview the driver about the harvest. Jill does her best to put them off but they are very persistent. In the end she directs them to the field, pops the tart in the oven and goes to warn Phil. He isn't up at the grain store, so she slips round to the field in the car. To her surprise Phil doesn't seem to mind – a sure sign that everything is going well. 'OK. David's on the combine. He can sort that one out,' he says cheerfully. Neil's baling in the next field, with Ruth squaring up the bales ready for carting, and Bert's hovering near the combine with the grain trailer. On the way back, Jill decides on the spur of the moment to go and check the cow in The Grove. When she gets back the tart is burned beyond redemption. Resignedly, she slides it off the plate into the bin and begins again.

Everyone is in high spirits when they come in for lunch. The barley's coming off well and doesn't need drying. Ruth talks to Jill while Phil and David go into the sitting room to watch the weather forecast on television. Jill isn't very pleased when she goes to call them for lunch as they've spread the straw and chaff all over her newly vacuumed carpet, not to mention the chairs. She knows better than to lose her temper. After all, it *is* harvest.

'How's the cow?' asks Phil.

'She's started,' replies Jill, dishing up. 'She's down by that big willow. There's a head showing. And one leg.'

'*One* leg?' asks David. 'Sure there weren't two?'

'I think it was only one. I'll go back again after lunch and have another look.'

'Well, it's important, Mum. Two legs and she'll probably do it on her own. One leg means trouble.'

'And we don't want trouble today if we can help it,' adds Phil.

But they soon have it. Jill's hardly stacked the dishes in the machine after lunch when David comes tearing into the yard in the Land Rover. 'Mum, can you rush into Borchester and get us a spare part?'

'What, now? This instant? I'm having trouble with the dishwasher – I think it's going on the blink.'

'Look, Mum, we're at a standstill out there.'

'Have you brought the part?' demands Jill.

'No, Bert's still trying to get it out.'

'Well, I'm not going, then. You know what happened last time. They gave me the wrong one and I had to go back again.'

'Please, Mum.'

'Oh, all right – but only if you look it up in the parts book and ring the dealers and make sure they've got it, first.'

Jill gets back from her mercy dash to find a young woman in the yard. She's the milk recorder and she wants the recording sheets so that she can get ready to record the evening milking. Jill wastes a quarter of an hour searching for the 'green sheets' in the dairy and the farm office until she's rescued by Phil who has decided to start the milking early, while they are still working on the combine. Meanwhile, can she take tea out to the others, please?

Everyone comes in at about 6.15 pm for ham salad and the re-made gooseberry tart, washed down by plenty of tea, after which Phil goes to have a look at the cow. Jill's right, there is only one leg coming and she'll definitely need help. With Neil gone to do the Hollowtree pigs and his own stock and the combine going well again, Phil decides to send for the vet and let him cope. He asks Jill if she could telephone the vet and take him down to the cow. 'He'll probably want to bring her up here, but you can give him a hand, can't you, love?'

At 9.30 pm Phil comes in. It's going really well and David wants to carry on after dark. Is that all right?

'Of course it's all right, darling. Martin's been and calved the cow. It's a nice little heifer, quite small and very white like the mother. He's left them both in the box. The electrician rang about the grain-dryer. He's coming out in the morning, about 10 o'clock. Tony called in – wants to know if you'll be able to cut his barley, when you've finished ours. There's plenty of coffee made and a big plate of sandwiches. And now I'm off to bed. Is that all right?'

To livestock farmers like Phil and David, straw is a vital by-product of cereal growing, used in large quantities for both feeding and bedding. To an organic farmer like Tony Archer it's equally indispensable; he feeds it and litters the stock with it, and it's also the basic ingredient of the farmyard manure which is his chief source of fertility. But, to a mainly arable farmer like Brian, straw is an unmitigated nuisance, something to be got rid of as cheaply and quickly as possible.

At Brookfield they bale all their straw using the conventional baler, turning out rectangular bales which are much easier to handle when bedding small yards and calf pens than the big round ones. Obviously, priority is always given to harvesting the corn in good condition, but they will do their very best to bale and cart the straw before it rains. If there's labour available the baler will be following the combine round the field. It draws a sledge behind it which enables the tractor driver to drop the bales off in bunches of eight. They can then be picked up eight at a time by a special loader mounted on the front of the tractor and placed straight on to a trailer. Back at the buildings they are stacked in the Dutch barn, with the help of an elevator. The wheat straw is used exclusively for bedding but some of the oat and barley straw, being more digestible, is fed to cattle. The bales themselves are very useful things to have around on the average farm and are used for all sorts of jobs, like building temporary pens for lambing and insulating buildings.

Brian, on the other hand, can bale as much straw as he needs for his

stock off the first two or three fields he combines. The rest is surplus to
requirements. He usually lets Tony come and bale some on the strict
understanding that the bales are carted straight away, and he lets Neil
have some for the winter in exchange for pig muck which he spreads at
Home Farm. Left to his own devices Brian would burn the rest of it
using the 'Bryant and May baler' as it's called. Unfortunately for him,
straw burning has become an unpopular practice with the general public
who have seen hedges and trees burned, smoke drift dangerously across
roads and their washing spoiled. Colonel Danby ran his car into the
ditch a few years ago as smoke from Brian's straw burning blotted out
his vision, but his complaints were as nothing compared with Jennifer's
when her swimming party at Home Farm was showered with black
smuts one sunny afternoon when the wind changed.

Straw burning still goes on but is now governed by an NFU code of
practice backed up by local by-laws. Only 25 acres can be burned at a

No time is lost nowadays after harvest before the cultivator moves in. The stubble
burning which has been carried out here leaves the ground in an excellent state for
working.

time with a wide cultivated strip around it to prevent spread. Burning is not supposed to take place at night or over weekends and Bank Holidays. The maximum fine for contravention is £2000. Burning has become popular not only as a means of getting rid of the straw, but also because it destroys weed seeds and disease spores and leaves the land in a perfect state of cultivation. Although Phil doesn't burn straw at Brookfield, he does sometimes burn the stubble, for the same reasons. If Brian could afford to wait he could sell his straw to Welsh farmers who are always short of it. But he prefers to forego the cash and get on with preparations for the next crop. Some of the straw he burns but an increasing amount now is being ploughed in. According to the pundits it doesn't increase the fertility of his land, but at least it gets rid of it without offending the locals.

In the old days before the combine harvester the straw *had* to be carted off the fields because it still had the ears attached to it. Back at the farm the sheaves were threshed during the winter and most of the straw turned into muck because the majority of arable farms then yarded beef cattle over the winter. Now, the combine does the threshing and spews the straw out behind it in endless swathes and, to make matters worse, there is much more corn being grown and the crops are much heavier, producing even more straw. And, these days, many arable farms carry no livestock to utilise it. All this adds up to the fact that about half the straw we grow — about 6 million tons a year — is surplus to requirements. All sorts of ways of using it have been devised both on and off the farm, but all together these outlets do very little to dispose of the surplus. The main limiting factor is that, because of its bulk, transport and storage costs are enormous, and anyway the cost of processing sometimes exceeds the value of the product. So, in the foreseeable future, straw will continue to be burned in large quantities in the field, with an increasing amount being incorporated back into the soil.

But at Brookfield, it's all lugged back to the buildings and Phil insists that every available space is filled with it. The fortnightly spraying of the potatoes to prevent blight, started in July, continues throughout the month. At some stage Phil will find time to go through his ewes and sort out those he intends to cull. And if the grass had not grown enough last month to take the second cut of silage this job too has to be fitted in between bouts of harvesting. The British weather usually obliges by providing plenty of periods when it's not possible to combine.

HOME FARM

With three combine harvesters, Brian can make big inroads into his harvest during good weather. He needs to be well mechanised since he has nearly 1200 acres to harvest and will be handling well over 3000 tons of cereals, oilseed rape and peas. His staff, augmented by a couple of students, are expected to work long hours during harvest and Brian will be driving round checking that everything is going according to plan. As soon as the weather prevents combining, the men are shifted to other harvest-related jobs – baling, carting or burning straw, machinery maintenance and repairs, ploughing, cultivating and spraying stubbles. There may also be a field to be put down to grass. Town friends holidaying with the Aldridges were amazed when Brian announced that one of his chaps would be sowing grass seed. They had assumed that grass was something which mysteriously grew on its own, and were surprised to find that it involved careful seedbed preparation and an expensive seed mixture. With only a limited amount of livestock Brian cannot practise the 'alternate husbandry' system favoured by Phil Archer, but he does plough up a grass field from time to time and take it into the arable rotation, and then he needs to sow another to replace it. August is a good time to get a good 'take' of seeds.

At Brookfield, where the acreages in grassland and crops are much more evenly balanced, it's possible to 'take the plough round the farm' with benefit to both. At one time rotations were very rigid; they had been worked out carefully to allow control of weeds and diseases and maintenance of fertility. The old Norfolk four-course system of roots-barley-seeds-wheat was a classic example. But once fertility could be bought in a bag of chemicals and sprays were developed to deal with most weeds and diseases, rotations tended to become much less circumscribed. Brian makes sure that he doesn't grow his sugar beet or oilseed rape on the same field two years running, but he has grown wheat and barley year after year on the same patch and continued to get good yields, admittedly at a high cost in sprays and fertiliser.

Giant round bales speed up the operation after harvest and tend to shed the rain. But smaller farms find them difficult to handle.

Grass is a wonderful builder-up of fertility. The fibrous roots of the grasses and clovers penetrate every part of the top-soil, increasing the organic content and restoring soil structure where this has been damaged by excessive cropping. The nitrogen produced by the clover roots together with the phosphate and potash applied in fertilisers over the years result in an explosive release of fertility when the land is ploughed. If the advantages of 'alternate husbandry' are so great it may be wondered why Brian also does not have half his farm down to grass. The reason is that grass can only be 'cashed' through livestock, and over the last 30 years or so arable crops have paid better than livestock. The exception is dairy cows, but Brian has never seen himself as a milk producer. The future now looks less rosy for arable farming and Brian may be forced to re-think his farming policy.

Brian does have one large-scale livestock enterprise, of course, in his flock of 600 ewes. Two hundred of these go to the tup this month – usually on 12 August. These are the ones which will begin lambing early in January. One day when it's not possible to combine, Brian and Sammy Whipple go through the remaining ewes to decide which ones are to be disposed of. Each year about a quarter of the flock is culled and replaced with young ewes. The main reasons for getting rid of ewes are that they have lost too many teeth, that they have had mastitis or that they are likely to cause trouble in the future, through illness, weakness or temperament. As the ewes file through the race, Brian 'mouths' each one, exposing its teeth. An adult ewe develops eight permanent front teeth (to replace the temporary ones it grows as a lamb); these are all on the lower palate, there are no incisors on the upper palate. The teeth arrive in pairs, the two centre ones in its first year, a further pair about 6 months later, the third within the next 12 months and finally the fourth outside pair. The ewes are known as 'two-tooth', 'four-tooth', 'six-tooth' and finally 'full-mouthed', which means that they are about four years old. Since breed and environment can affect the rate at which teeth develop, some farmers prefer to age a ewe by the number of crops of lambs she has had. After a ewe is full-mouthed she begins to lose teeth, and it's at this stage that Brian sells them. As he mouths each ewe, Sammy Whipple reaches underneath and feels the udder.

While they work they are closely watched by Sue, a friend of Debbie's who is staying; she is fascinated by the farm and wants to know more.

'Why does it matter if they've lost a tooth or two, Mr Aldridge?'

Brian marks a ewe on the head and lets her go. 'Well, they can't eat so well, and they live by eating and it's not as if they'll grow any more teeth. We call them broken-mouthed when they get like that.'

'They should fit them with false teeth, then,' she giggles.

Brian's mouthing another ewe. 'Actually they've tried that, Sue, but I don't think it caught on commercially.'

A ewe's age can be gauged roughly by 'mouthing' her – counting the front teeth which grow only on the lower palate. A 'full-mouthed' ewe has eight teeth. This one has lost four and is 'broken-mouthed'. Farmers like Brian and Phil replace ewes at this stage with young ones.

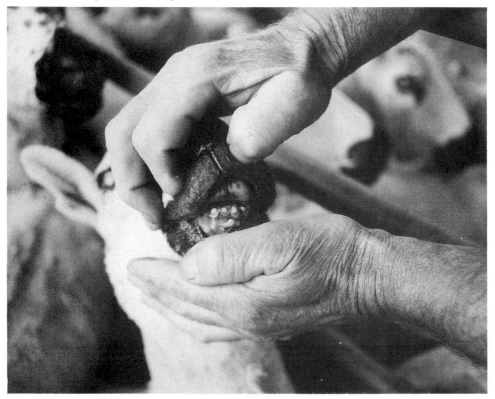

Sue leans over the race. 'Why do you mark some on the head and some on the back, Mr Aldridge?'

'The ones I mark on the head are fit to breed from – they've just lost some teeth. There's nothing wrong with their udders. Sammy's the expert on the other end. If he doesn't like the feel of their bags it means they've got mastitis and out they go, pdq, like this one,' he adds, giving it a dab of red on its back.

'Out where?' Sue persists.

'To the slaughterhouse, I'm afraid, dear. Can't pass them on. Suppose she has twins next year and only one quarter to suckle them on – no good to anyone.' Brian grabs another ewe. 'Now here's one with really good teeth. Look, Sue,' he says, holding its bottom lip down to expose eight perfect incisors. 'What's she like behind?'

'Fine,' grins Sammy, and Brian marks it on the head.

'Why are you marking it then, Mr Aldridge, if it's all right?'

'She's fine in the mouth and the bag,' he explains. 'But she's only brought singles for the last three years, and there's no profit in singles.'

Sue gasps. 'Do you know all the sheep then?'

'No, but Sammy does. He's got this little book with their ear numbers in, and if they don't perform properly, he puts a black mark against them.'

Out of his 400 ewes, nearly 90 end up with a mark on their heads or their backs. Some, Brian knows, would be all right to keep for another year, but by culling annually on the same basis he preserves the balance in his flock – a mixture of ages – and doesn't suddenly have to go out and replace half of them at once. Some farmers rely on buying broken-mouthed ewes 'guaranteed in the udder' and will carry on breeding from them until they haven't a tooth in their heads. Brian prefers to leave less to chance.

One damp, overcast day at the end of the month, when there is obviously no chance of harvesting, Brian decides to remove the antlers from his two breeding stags. He does this to prevent fighting and injury during the rut. The 'velvet' – skin which covers and nourishes the antlers while they are growing – has now disappeared. Once the antlers are fully grown, the blood supply to the velvet stops and the stag 'frays' his antlers against trees and bushes to get rid of it. By the time Brian removes them, the antlers are mature and burnished bone.

He has to tranquillise the stags in the same way as he did the hind which had calving difficulties in June. Once they are 'out' it is a very quick job to take off the antlers with a hacksaw just above the pedicle. The operation does not hurt the stag which gets shakily to its feet a few minutes later, shaking its head and no doubt wondering where its 6 or 7 kg of antler has gone. The antlers are in demand for a variety of uses, such as handles for knives and walking sticks; an average set would fetch Brian between £30 and £40.

BRIDGE FARM

No time is easy on a small farm where the boss and his wife constitute the major part of the labour force, and some small farmers go for years without a break. But Tony and Pat sometimes take a short holiday at the beginning of August. They probably wouldn't contemplate it except for the children. But with a lot of the cows dried off and the wheat and barley not yet ready to harvest, it's as good a time as any. Sometimes, however, the hassle of rushing to get jobs done before leaving, finding and paying someone dependable to cope while they're away and catching up on the work when they return is too much to face, and they settle for taking the children on day-trips. The situation has become more difficult since they became fully organic with the added complication of making and distributing yoghurt. They have also had other troubles. Mike Tucker earned himself black marks for treating one of their cows with antibiotics not long ago, firmly against their principles. They also know that if they decide not to go they certainly would not find themselves short of a job. The weeds continue to flourish even though it's holiday-time, and Tony would be taking the tractor hoe through the cabbages and swedes and gently earthing up the leeks, leaving Steve to pull out the worst of the weeds in the rows. He'd also be baling some of Brian's straw and lugging it back to his buildings for feeding or bedding during the winter. Brian wouldn't charge him for it but would be in his usual hurry to get it cleared. 'Look,' he barks on the telephone, 'you can bale as much straw in that field by the road as you want, but I need it clear by the end of the week. I'm cultivating it on

Monday and if there's any straw left there then I shall burn it. OK?'
Tony puts the phone down and sighs, thinking of a sun-drenched beach
in Greece.

A constant source of worry to Pat, especially at this time of year is the
risk of accidents to her children and their friends. Farms are the most
attractive and dangerous playgrounds. This is particularly true of a
place like Bridge Farm with its old buildings overflowing with hay,
straw and baler twine, its lofts, missing floorboards, broken ladders and
old machinery. It's a perfect place for hide and seek or something more
exciting – with potentially lethal hazards lurking round every corner.
Each summer produces a catalogue of fatal accidents on farms, and Pat
spends a lot of time checking on the children's whereabouts. One
morning the children sit wide-eyed round the breakfast table as she
reads to them from one of the farming papers.

'Now, children, you know I'm always warning you about you and
your friends playing round the buildings – listen to this. It's a list of
terrible accidents to children on farms. Tommy, a child of your age
climbed a heavy gate leaning against a wall; it fell over on top of him
and crushed him. Tony, there's a gate leaning against the parlour wall,
isn't there?'

'Yes, but I've tied the top of it to the post,' he replies.

'A 12-year-old tried to get on to a moving trailer and the wheel ran
over him and killed him. Ooh, and a ten-year-old was hanged with baler
twine, playing in the buildings. You must never play around with baler
twine like that.' Pat looks at the report closely. 'Oh, Tony, a baby run
over by a reversing tractor, isn't it awful? And a three-year-old drowned
in a farm pond. We ought to fence that pond, Tony,' she continues, 'and
a girl of Helen's age run over by a lorry delivering feed'.

'But, Dad's always trying to get me to help on the farm,' John pipes
up. 'I know it's dangerous.'

'It's all right if you do as you're told,' growls Tony. 'Now come and
give me a hand with these bales. And the rest of you – be careful.'

Tony and Steve, the YTS lad, are pulling up ragwort plants and stuffing
them into old feedingstuff bags. Ragwort is a tall handsome plant with

an attractive yellow flower, now in full bloom. Unfortunately it's also highly poisonous to livestock. Stock will normally avoid it if there's plenty of grass, so it's left to seed undisturbed and consequently spreads. The main danger to Tony's cattle lies in its being incorporated into hay or silage, where they can't easily discriminate but where it retains its lethal properties. It attacks the liver and usually results in death, although this may not occur for several weeks or even months. This is the plant which Joe Grundy was destroying when surprised by the meddlesome Mrs Snell. Ragwort is one of three types of injurious weed listed under the Weeds Act 1959 – the others are thistles and docks. The Ministry can require an occupier of land to deal with these weeds, and if he fails to do so can have them dealt with and recover the cost. Prosecutions are rare, however; most offenders submit to persuasion, while sensible farmers get rid of their ragwort before it can cause harm.

Paradoxically, there is now a list of about 70 wild flowers which, under the Wildlife and Countryside Act 1981, it is an offence to 'pick, uproot, sell or destroy'. A farmer who refused to control his ragwort could, if convicted, be fined £75. But anyone insensitive enough to destroy a monkey orchid could be fined a maximum of £2000.

September

Marking the end of the corn harvest and the beginning of autumn, September ushers in a glorious mixture of seasonal offerings, among them blackberries, evening mists, hazel-nuts (where the squirrels have not reached them first), lavender, mushrooms and glistening spiders' webs carpeting the grass in early mornings. The year finally drops her mask of eternal youth, sustained with increasing difficulty during August, and submits gracefully to full-blown middle age. The rustle of spring in the trees gives way to a drier, more brittle susurration as the leaves begin to loosen; not all of the trees have begun to change colour yet, although the virginia creeper on Brookfield farmhouse is a blaze of red. It's a time of change. As the Ambridge cricket team dons its white flannels for the last time, the Horobin youngsters are already kicking a football about on the village green. Tom Forrest gratefully soaks up the afternoon sun as he gathers his plums, but in the cool of a September evening he's pleased to warm his legs in front of a good log fire.

For many the big event of the month is the Ambridge Flower and Produce Show. Age-old rivalries are rekindled as the exhibits are laid out on the benches: runner beans 18 inches long, onions as big as croquet balls, lettuces the size

A sale of Suffolk rams. This is the breed which they use at Brookfield on their flock of 'mule' ewes to produce good lambs for the butcher. Phil usually buys one or two replacements in early autumn at Hereford market.

of cabbages. Accusations and counter-accusations are rife, although these are more often retailed through a third party than face to face. The old anecdotes of past shows are regurgitated in The Bull of an evening, losing nothing in the telling as the talk moves from Joe Blower's magic way with leeks to Walter Gabriel's exploding marrow.

It's not only the fruit and vegetables which are open to contest. There are the flowers – classes for a wide range from 'Dahlias, five blooms, any variety' to 'Gladioli, three spikes' – and the floral art section featuring this year as its dual central challenge 'An exhibit in clashing colours' and 'Tranquillity – an exhibit including water'. Jill and Pat are exhibiting free-range eggs, Prue Forrest has entered a whole kitchen cupboard of preserves, including this year's crab apple jelly which has come out a lovely dark shade of pink, Martha Woodford has put in some of her green bean chutney and Mrs Antrobus, who has been quietly experimenting at Nightingale Farm, has entered her first attempt at elderflower wine. There is much good-natured chatter as the exhibits are laid out, followed by an animated post-mortem when the tent is open for inspection following the judging; this is echoed and augmented over the tea-making at the village hall during the next WI meeting.

Mushrooms are not as common as they once were around Ambridge, probably due to the increased use of chemical fertilisers, although they still grow in one or two fields at Grange Farm. Joe, who had been nurturing a little colony up by the road, watched outraged as Kevin the tanker driver stopped his milk lorry alongside and calmly picked them into a polythene bag. It almost brought on an attack of Joe's farmer's lung, although there was nothing he could do about it as no one owns wild mushrooms, a fact of which Kevin would have been quick to remind him had he objected. After that he took to disguising any he found, which were not ready to pick, by covering them with grass, especially a collection of horse mushrooms which he loves Clarrie to cook for him in a casserole with bacon and tomato.

For Joe – and for Tony Archer – some of the delights of the season, like mushrooms, and blackberry and apple pie, are overcast by the knowledge that 29 September is Michaelmas Day, when a further half-year's rent is due.

BROOKFIELD

The cereal harvest is normally over by the second week of September, sometimes earlier, according to the weather. Once the combine is put away everyone can concentrate on clearing the remaining straw and cultivating the stubbles. The oilseed rape is drilled around the first week of the month and is showing through by the end. The protective spraying of the potatoes continues. There is plenty to do but also time to relax a little. Phil has been bringing on a couple of gilts out of Freda for the Rare Breeds Sale at Stoneleigh. He had intended to exhibit them

The 'conventional' baler, still favoured at Brookfield where they use the bales not only for feeding and bedding but for jobs like building lambing pens and lining the potato store to keep out frost. The twine which binds them also has many uses.

as well but an established Middle White breeder came over to have a look at them last month. 'They're nice gilts, Mr Archer. But they're – how can I put it – not quite up to the mark, are they? Not for showing anyway, that's my advice.' The chap had been at the game much longer than Phil so, suitably chastened, he decided to leave it for this year.

The decision turns out to be fortuitous in one way since they are busy combining wheat at Brookfield on the first day of the event and David, who has never been very keen on his father's apparent obsession with Freda and her offspring, would no doubt have viewed his absence as verging on the irresponsible. Once the field is finished and the combine sheeted up, Phil sets about washing his pigs with the help of Kenton who is home for the weekend and has offered to accompany him. He beds them deep in straw and hopes they won't get dirty overnight.

Judging Longhorns at the show and sale of rare breeds. The setting up of the Rare Breeds Survival Trust has halted the extinction of breeds of traditional farm livestock. Between 1900 and 1973 at least 20 breeds ceased to exist.

By 9 o'clock the following morning, Saturday, they are penned in the huge cattle shed at the National Agricultural Centre, having been inspected by the veterinary surgeon and numbered across the shoulders. With time to spare Phil and Kenton wander off to look at the other Middle Whites and then the rest of the pigs: Gloucester Old Spots, Saddlebacks, British Lops, Berkshires (a black pig with the same dished face as the Middle White), Large Blacks and Tamworths, (the golden red breed once popular in the Midlands). On the blackboards above the pens owners have scrawled information not given in the catalogue, such as 'Scanned and warranted in pig'. They are hailed by the breeder who advised Phil not to enter his gilts for the show on this occasion. He takes them along to see his young boar which has won the breed championship; after inspecting all the rosettes adorning its pen Phil feels that his advice was probably sound.

Since there are still a couple of hours before the pigs are sold, Phil and Kenton go to see the cattle and sheep. Phil is amazed at the variety of people thronging the alleyways between the stalls and pens. Landowners rub shoulders with smallholders, punk hairstyles with shaggy beards, babes in prams and old men with sticks. There's a queue to join the Rare Breeds Survival Trust which organises the annual event. Phil and Kenton experience a rich intermixture of smells as they move from the pigs through the cattle to the sheep and, finally, the goats (which, fortunately for the olfactory senses, are relatively few in number).

The cattle vary in size from the huge Longhorns with their massive sweep of curved horn, to the little Dexters, some of them no taller than a large dog. The ancient British White, with its characteristic black nose, ears and feet, has always been a rare breed to Phil, but he can remember when several of the others like the Beef Shorthorn were anything but rare. He remembers his father taking him to see a commercial herd of Gloucester cattle, the attractive red breed with the white stripe running down the back, hindquarters and tail; their rich milk, he tells Kenton, was being turned into Double Gloucester cheese.

The sheep are as varied as the crowds gathered round them. Little dark brown Soay lambs not much bigger than large cats, woolly Cotswolds almost as tall as donkeys, a Portland ram with three twists to its heavy horn, a Lincoln Longwool ewe which will yield 20 lbs of wool. With so many to sell, two auctioneers are already at work. Phil bumps into an old friend who seems very much at home there as he smiles and waves at people pushing past them in the narrow alleyway.

'There are far more folk here than I expected,' says Phil. 'This rare breed thing seems to have taken off in a big way.'

His friend looks round. 'Yes,' he replies, somewhat doubtfully. 'The trouble is that not all of them know too much about breeding.'

'Still, at least the breeds aren't dying out,' Phil comments. 'Anyway we must go back to the pigs – they'll be selling them soon.'

They find Phil's breeder friend rubbing wood flour (very fine sawdust) into the coats of his two gilts. 'I've sponged them and rubbed this in,' he tells them. 'Brush it out just before you take them into the sale ring – you'll find it'll give them a lovely sheen. You'll have to learn the tricks of the trade if you're going to start showing pigs.'

Eventually it's their turn in the ring. Phil has borrowed a white coat and looks very professional as he nudges the first of the gilts round the small ring, with a couple of hundred people, some of them potential buyers, looking on. He's not feeling too optimistic as the previous lot was unsold. The auctioneer, having worked himself up into a frenzy of excitement ended lamely, as the bidding stuck at 115 guineas, 'I'm sorry, sir, afraid I can't do it at that price.' But he immediately brightens up again at the thought of another challenge. 'Lot 152 comes from Mr Philip Archer . . .' The bidding starts at 50 guineas and the hammer falls at 95. The second gilt makes 85 guineas (the selling is traditionally carried out in guineas – 105 pence – the auctioneers keeping the odd 5p as his commission).

'What did you think of the prices, Dad?' Kenton asks cautiously as they drive home.

'Not as much as I'd hoped,' his father replies. 'Still, it's been an experience. I'd have liked it to be nearer 120 guineas. There's obviously more to this business than I thought. Perhaps we'd be better off selling them as porkers to that chum of yours in Cheltenham. Anyway, it'll give David something to pull my leg about.'

The autumn calving is in full swing at Brookfield and there will be 30 or 40 cows due in September. Most will manage on their own; perhaps 10 per cent will need assistance and one or two may require the services of the vet during the 3-month calving period. It's a time of great anxiety for everyone involved; it's difficult to decide when to intervene if a cow is taking a long time to calve and whether the operation is too complicated for the farm staff to handle. A dead cow can mean a loss of more than £600 with as much again in lost milk over the lactation. A dead heifer calf might have become the herd's top yielder; a dead bull calf could have fetched £200 at Borchester market after a few weeks.

The cows nearest to calving are in Pikey Piece, the little field on the corner near the farm, and Phil is fetching four which are due in the next few days. Although there is nothing nicer than to go out and find a cow licking a steaming new-born calf which has arrived unexpectedly without complications, Phil prefers to have them back at the farm where they can calve under supervision. The first thing which any experienced stockman does as he enters a field is to count the stock, and Phil's first intimation that something might be wrong is when he sees only 18 cows where there should be 19. In such a small field it doesn't take him long to find the reason for the discrepancy; no. 439 is lying on her side breathing heavily with her head twisted back. Phil recognises the classic signs of milk fever, but what alarms him is that the cow is lying on a steep bank and only about 8 feet from a narrow ditch. One more roll and she could be into the ditch, blocking the flow of the water and in danger of drowning. He has to think quickly. He begins by pulling her neck to its proper position, but the moment it's released it springs back. This confirms in his mind that it's milk fever. For the next 10 minutes, using every ounce of his strength, he works to move her a few feet to a point where she is wedged against the trunk of a willow. He twists her head, he pulls her tail, he folds her legs and tries unsuccessfully to turn her, he braces his back against her and pushes with his legs. It's not easy; he's trying to move 13 or 14 cwt of flesh but at last he manages to manoeuvre her into a safer position. Only then does he trot, showing a surprising turn of speed for a sexagenarian, to an adjoining field and bring back a bale of straw to wedge her even more securely.

Milk fever is an ever-present threat to the dairy herd and can prove fatal; it would be very unusual to get through a calving period without half a dozen cases. It's the result of a shortage of calcium in the cow's blood which in turn is caused by the demands made on the available calcium by first the growing foetus and then the production of milk. It normally occurs shortly after calving but sometimes, as here, it can affect the cow beforehand. The treatment, which is fairly simple but dramatic in its effect, consists of administering calcium borogluconate by injection either under the skin or, in emergency, into a vein.

Phil has just decided that it's safe to leave the cow while he fetches the necessary equipment when he hears a shout from the gateway. It's Ruth who had been waiting back in the yard to help him put the cows in the loose boxes. She gladly offers to race back and collect the equipment while Phil keeps an eye on the cow. Before long she's back in the Land Rover and hauling out a 2 gallon can of hot water and a couple of udder cloths.

'What on earth's that for?' Phil enquires.

'You dip the cloths in hot water and hold them against the cow where we're injecting. Increases blood circulation and carries the injection away quicker,' she explains, proudly. 'I learned about it at college.'

Phil smiles. 'Well, let's put it to the test,' he says, kindly. He inserts the needle in the cow's shoulder. While they wait for the injection to take effect, Phil makes a mental note to fence off this bit of the brook to avoid further trouble. In about 20 minutes the cow tries to lurch to her feet, nearly knocking Ruth over in the process, and 10 minutes later she's on her feet, still a little shaky. Ruth comes back with the stock trailer and they load the animal to take her back to the buildings. The cow calves normally the following day without further trouble.

HOME FARM

With a much bigger arable acreage than Phil Archer, Brian – or at least Brian's labour force – tends to be working flat out during September. It's usual for him to have some of his barley drilled by the end of the month; and he spends a lot of his time keeping the wheels turning, including keeping a close eye on activities at the grain store, but he would certainly not turn down an invitation to shoot, either grouse or partridge.

His own stock of partridge has now increased to the point where he can resume shooting; the season officially opens on 1 September. Before the last war a large arable farm like Brian's would have been capable of holding perhaps 40 or 50 pairs of partridge, each rearing up to a dozen chicks in a good season and providing several days' shooting without jeopardising the stock. Unfortunately the arable farming of the last 30 years has militated against the survival of the partridge and numbers were brought to a very low level. Two practices have been blamed for the decline – the removal of hedges, and chemical spraying. Partridge thrive in hedgerows, the rougher the better. Bigger tackle encouraged farmers to make fields larger, and where hedges were left they were trimmed and sprayed down to mere skeletons of their former selves. With no hedges providing decent cover there was nowhere for the partridge to nest. Spraying of the cereal crops killed off the insects and weed seeds on which the birds feed so that when a partridge managed to

Loading a modern pneumatic drill, capable of sowing 80 acres a day. Brian Aldridge recently acquired a machine of this sort which handles only the seed corn, unlike Phil Archer's combine drill which applies seed and fertiliser at once.

hatch a brood of chicks there was nothing for them to eat. It took some years for farmers to realise what they had done, and some of them, like Brian, are now taking active steps to encourage the birds.

But it's not only partridge that Brian is shooting this month. He's also beginning to 'harvest' his deer – the stags not needed for breeding, those about 15 months old. Although deer can be slaughtered in abattoirs Brian, like many producers, prefers to kill his own on the farm. He simply sprinkles some nuts on the ground in the paddock and when the deer come to feed he shoots one with his rifle at fairly close range. Although it sounds a bit callous, he argues that it's the kindest way. The selected deer is killed instantly while the others appear supremely indifferent to its fate and carry on eating their nuts as if nothing had happened. After bleeding the carcass Brian usually takes it to a butcher who dresses it for him.

What happens to it after that depends on the market. It may go to a hotel, like Grey Gables, it may be retailed through a game dealer or it may be taken back to Home Farm for sale privately from the freezer. If Brian has too many deer to dispose of locally he sells them through the British Deer Producers' Society – a co-operative set up by deer farmers to market venison. With the accent these days on healthy eating, venison has a great deal in its favour, containing considerably fewer calories, less fat and more protein than chicken, veal, lamb, pork or beef, though at the moment it's much more expensive than other red meats.

Some time during September, once the corn is all cut, the foxhounds will be cub-hunting at Home Farm. Cubbing is looked on by some of the non-hunting fraternity as an activity akin to clubbing young seals to death, whereas the afficionados would point out that the so-called cubs are well-grown and have probably already killed their first pheasant or chicken. The aim, they would explain, is to thin out and disperse the cubs which are still in family groups and to teach the young hounds that their sole permitted quarry is foxes.

At this time of year most of the foxes are living above ground – first in the corn, and later, when that has been combined, in kale, root crops or woods and other thick cover. Later in the season one of the worst crimes in the hunting field is to head a fox (that is, cause it to change direction) but during cubbing this is not only allowed but encouraged. The field – those riding, augmented by a crowd of foot followers – spreads out round a covert or field of roots while the hounds draw it. When a fox tries to break cover the aim is to turn him back so that the hounds can carry on hunting him.

Hounds meet at this stage of the season at about 7 am and stay out until perhaps 11 am, depending on the scent. If the meet is at the end of Brian's drive, it's surprising how many inhabitants of Ambridge find it possible to be there, even if it's only for an hour or two. Shula's there on one of Chris's horses with Caroline, both looking very smart in 'ratcatcher' (the 'uniform' for cubbing of hacking jacket, shirt and tie). 'I can't stay out too long,' says Shula, 'I'm meeting the farm manager at the estate office at 10 o'clock.' To which Caroline replies, 'Neither can I – Mr Woolley doesn't even know I'm here.' Tom Forrest is there, with his van; although he has never worn a hunting coat in his life, he'll see more of the hunt than anyone else. George Barford is also there; it will give him an excuse to have a good look round Brian's shoot on the pretext of enjoying a morning's sport. Also there, with an ulterior motive, is Eddie Grundy who has persuaded his father to do the milking while he explores the prospect of a little poaching. Chris is out on a new horse she wants to try. And Bert Fry has asked Phil Archer if it's all right if he comes in late and 'makes up for it some other time'. There are about 15 mounted and twice that number on foot as they move off to draw the long covert beside the Felpersham Road. The sun has already broken through and will soon disperse the thin mist and herald a really fine September day.

Brian, who met them at the end of his drive to tell them where they might find a fox and which parts of the farm he would prefer them to avoid, has now gone back to the yard to sort out the day's work. At the moment he's more concerned with getting some barley drilled than catching foxes. Like other arable farmers, he's not keen on having the hunt over his land although he would regard it as unsporting and anti-social to refuse. He's quite glad to see the foxes thinned out before they take too many of his pheasants, but he and his keeper know of more certain ways of reducing their number. In The Bull the other night, Brian was actually heard defending hunting against a vicious attack from one of the newcomers from the Brookfield barn conversions. 'Look,' he said, 'no one can pretend that hunting's the most efficient way of getting rid of foxes. I could destroy every fox on my land in a very short time by shooting,' he paused, 'and other means I'd rather not go into in this company. But do we want a countryside without any foxes?' No one answered, so he continued, 'On the other hand foxes don't have any natural predators and could over-run the country if they weren't controlled. What hunting does is preserve a balance between too many foxes – and no foxes at all.'

BRIDGE FARM

Tony and Steve, are treating a small bunch of yearling bullocks affected with New Forest disease. This is an affliction of the eye which occurs most often during the summer and usually in younger cattle. It starts off with the eye watering which unless checked can lead to an ulcer. The disease is thought to be spread by flies and obviously causes extreme discomfort. Normally Tony would bring affected cattle indoors out of the bright sunlight and away from their fellows to prevent spread. But these yearlings are in a field away from the farm near the village, and it's difficult to house them. So he's hoping, having discovered the outbreak in its early stages, that he can contain it without having to bring them back to the buildings.

They have constructed a makeshift pen adjoining the gate into which they have crammed all the bullocks; the less room the animals have to move around the easier they are to handle. Tony grabs their heads, one by one, holding them as steadily as he can while Steve squirts an antibiotic powder into each of their eyes. There's plenty of shouting and mild swearing as the animal blinks its eye or jerks its head at the crucial moment – or the plastic 'squirter' refuses to squirt. Although only three are affected, Tony decides to treat them all. He'll repeat this for several days and if they get any worse he'll have to bring them home. While routine treatment with antibiotics is not allowed by the Soil Association it is permitted 'to prevent unnecessary suffering or where no other effective treatment is locally available'. Pat wanted him to try a homoeopathic cure but was persuaded that it was too urgent and they were too busy with other jobs to experiment. 'We'll try it another year,' Tony promised.

New Forest is not a killing disease but it can waste a great deal of time, a commodity in short supply at Bridge Farm this month. Tony and Pat are being pulled in several directions at once. They have started bunching carrots – pulling the larger ones and marketing them with their leaves on. They have to harvest their spring wheat and arrange for its sale, keeping some back for grinding into organic flour for local sale. The cabbages and swedes need the tractor hoe through them a couple of times and the leeks have to be earthed up again to blanch them. Straw must be baled and carted, and with the autumn calving under way Tony

A bad attack of potato blight which destroys the foliage, reduces the yield and can infect the tubers. As an organic farmer, unable to use protective sprays, Tony Archer has to face regular attacks; he just hopes they will not come too early, before the potatoes have bulked up.

never knows when he's going to be required to deal with a difficult birth. On top of this the cattle need milking twice a day and the yoghurt and butter must be made. They don't want additional trouble such as New Forest disease, and anyone in the village making sympathetic noises is likely to be invited to come and lend a hand. But Tony, somehow, still finds time of an evening to pop down to The Bull to let everyone know how busy he is.

It was against this active background that Pat almost lost her temper in the village shop one day when she rushed in to buy some bacon and found Lynda Snell complaining about the price of biscuits.

'Honestly,' she said, largely for Pat's benefit, 'there's so many surpluses, you'd think food would be cheaper, not dearer. Look at these, 65 pence! I swear they were only 50-something last time I bought them.'

Pat bridled, but said nothing.

Undeterred, Mrs Snell went on. 'You farmers must be doing *very* well at present.'

'Will that be all, my dear?' asked Martha Woodford, anxious to avoid trouble.

'It's all I can afford,' declared Mrs Snell pointedly.

This was too much for Pat. 'Actually,' she said, icily, 'most food is cheaper than it was ten years ago – in real terms, that is.'

'These biscuits certainly aren't,' said Mrs Snell, glad to have drawn a response.

'Well, it's not because farmers are earning more,' said Pat, trying to keep herself under control. 'The NFU produced some figures which showed that, with bread, the farmer is getting a much smaller percentage of the sale price these days – and I expect it's the same with biscuits.'

Mrs Snell began to put her purchases into a Harrods carrier bag.

'It's probably the manufacturers making the profit,' Martha intervened gently. 'Not the farmers.'

But Mrs Snell refused to give up. 'Robert was telling me that the Common Agricultural Policy costs the average family £13 extra a week in higher prices and taxes. He read it in the paper. So where's it all going, if you farmers aren't getting it, I'd like to know?'

'I 'spect it's all them Eurocrats or whatever they calls them, living it up in Brussels,' purred Martha, determined to keep the peace.

But Pat was beginning to lose patience. 'I don't know what your husband read in his newspaper,' she snapped. 'All I know is that people are having to work for *less* time to buy their food now than they did ten years ago. I saw the figures when they came out last winter. There was a whole lot of things mentioned – bread was one, *and* eggs, *and* pork.'

Mrs Snell seemed lost for a second and then bounced back, 'Well, you seem to be doing quite well with your stuff – butter at £1 for half a pound, did I hear?'

'That's special stuff and you know it,' she retorted. 'It's farmhouse made *and* organic. Ordinary butter's actually cheaper in real terms than it was ten years ago.'

'Can I get you something, my dear?' Martha urged.

Mrs Snell put her bag down as if squaring up for a long debate. 'Something'll have to be done about the CAP.'

'Mrs Snell,' said Pat, breathing heavily, 'I've got a hundred pots of yoghurt to make when I get back; I've got a calf which won't drink properly, we've got a cow with mastitis which has to be seen to every 2 hours, Tony wants me to go and help him pull carrots, I've got four beds to make and lunch to get and Steve's off on day-release. So why don't you go home and get on with your *petit point*?' She turned to face the counter, 'And now, Martha, may I have a pound of bacon, please – English for preference.'

Mrs Snell picked up her bag and left.

Living off the Land

Although only an hour's drive from Birmingham, Ambridge has remained surprisingly agricultural. True, commuters are moving in to places like Glebelands and the Brookfield barn conversions, but one way and another the farms and the estates, the fields and the woods still provide much of the employment and income in the village. Jack Woolley may attract city money to Grey Gables, Sid Perks may exploit the passing trade at The Bull and Chris Barford continue to draw the middle-class children from Borchester to her riding school, but the land remains at the heart of the life of the village.

However, not all of those who earn a living from the land are in the Range Rover class. There's Joe Grundy at Grange Farm, for instance, and his son Eddie and long-suffering wife Clarrie. It's a wonder how they've survived the twice-yearly rent demand for so long. But most parishes seem to include a family like the Grundys, not all as cantankerous or devious as Joe, perhaps, nor as workshy or roguish as his son, but families running holdings too small to make a decent living and not farming them particularly well. They survive for two reasons. The first is that they spend very little. If Eddie backs the trailer over the only muck fork and breaks the handle, Joe will spend a morning fashioning

another one out of a piece of ash cut from the hedge. (If the same thing happened at Brookfield, Phil would remonstrate with the offender and then spend £20 on a new one.) The second reason is that no one will allow them much credit. Joe has an absolute overdraft limit of £5000 at his bank and most of the firms with whom they do business try to extract a cheque before they unload. So they can't get deeply into debt which, as interest rates have risen, has been the undoing of many a farmer in recent years.

Grange Farm is 120 acres and rented from the Bellamy estate. In 1987 Joe was successful in getting a rent reduction of £3 an acre to £45 to compensate for falling margins and he and Eddie are hoping for a further cut at the next rent review in 1990. It's not the best of farms, being a little too banky for much arable, but the Grundys have palpably failed to make the best of it in the past. Things got so bad ten years ago that Andrew Sinclair, the agent, served a 'Notice to Remedy' on Joe, requiring him to deal with about 40 lapses in his tenancy agreement within 6 months: involving re-hanging gates, repairing fences, clearing ditches, cutting hedges and destroying weeds. The penalty for failing to meet his obligations was the risk of eviction from the farm. The situation has improved of late, brought about indirectly by the arrival of Eddie's two sons. Occasionally, in an expansive mood, he will take them to the top of the farm and wave his hand saying, 'One day all this will be yours.' The sad fact is that, in the absence of a miracle, he's wrong.

The farm carries a dairy herd of 40, mainly Friesians with a few Ayrshires bought because they were cheap. They grow about 50 acres of cereals, some of which they grind for feeding on the farm. They usually fatten about 50 turkeys for Christmas. There are a few beef cattle on the farm – they keep back some of the bull calves and sell them at various stages, depending on the cash flow or available grass or silage. From time to time they feed a couple of pigs, and last winter Joe ventured, somewhat disastrously, into sheep. Their last set of farm accounts showed a surplus of about £7000 for the year. To this can be added Joe's old age pension, Clarrie's earnings and Eddie's extra-farm activities. Small wonder that Clarrie was applying for Family Credit. Joe blames their present financial difficulties on the milk quota. 'Cut us off in midstream,' he explains to anyone who'll listen. 'We was set for big expansion. Up to 100 cows. Then them beggars in Brussels does this to us. Hamstrung we are. Stuck with 40 cows. What use is that? And my little grandchildren needing new shoes.' Of course it's all an illusion. He would never have built the herd up, and milk quotas, far from harming him, have helped by increasing the price he gets for his calves and cull cows.

The long-term future for the Grundy family looks grim. Joe's in his seventies and will probably carry on milking and complaining until he dies, although increasing age and his farmer's lung will diminish his contribution to the running of the farm. Meanwhile Clarrie has expectations for William and Edward which, as they grow up, will be difficult to sustain on the profit from Grange Farm. Eddie, now 37, will have to do more work both on and off the farm if they are to survive

although were he to take permanent employment away from the farm he would jeopardise his right to succeed to the tenancy. Clarrie may hold the key to survival in the short term. Once Edward is at school she could take a full-time job and Joe will just have to keep quiet if his tea isn't steaming on the table when he gets in from milking.

The best long-term hope is that Eddie can work on his erstwhile schoolfellow Lilian Bellamy, now his landlord, with whom he appears to have a 'special relationship', and persuade her to add the adjoining Ambridge Farm (at present farmed by the estate) to Grange Farm, including some extra milk quota – and then really get stuck in.

Mike Tucker, one imagines, was born to carry a chip on his shoulder. Nothing ever seems quite fair; there's nearly always something or ` someone worthy of resentment. He has always appeared to be at the centre of a disagreement or disappointment. Occasionally he will put his arm round his wife Betty and look at his two children and think how lucky he is; but not often. More frequently he's been heard railing at some injustice or other, the job he wasn't offered or the one he did for which he was underpaid.

It has become increasingly difficult over the years for a farmworker to become a farmer. Mike tried it and failed disastrously. Had he started ten years earlier or with twice as much capital he might have made a go of it but by the time he got into Ambridge Farm the squeeze on agricultural returns had begun and he 'over-borrowed'. Had the landlord-tenant partnership with Haydn Evans continued, Mike might have survived. This was an arrangement under which Haydn owned the holding and Mike farmed it, the proceeds being split between them. The trouble was that Mike wasn't generating enough 'proceeds' for Haydn's liking and so he gave notice to wind up the partnership in 1983, and sold the farm. Dan Archer persuaded Lilian Bellamy to let Ambridge Farm, previously run by the Pounds, to Mike; he'd always had a soft spot for him, ever since he was cowman at Brookfield. As things turned out it wasn't the good turn the old man intended. Within a couple of years Mike was running into serious trouble; his bank overdraft was mounting and he owed money in all directions, milk quotas had been imposed, he was selling off assets and his cheques started to bounce. In January 1986 he decided to declare himself bankrupt; when his affairs were settled he wasn't desperately in debt but he knew that he was fighting a losing battle and that things would only get worse.

Now, three years later, Mike finds himself back at Willow Farm, though this time tenant only of the farmhouse. He seems reluctant to face the future as a self-employed forestry and farmworker, preferring to re-live the past by chucking bales of hay to the Brookfield heifers or helping Neil feed his pigs. On the face of it his prospects are good. His forestry course should open the gates to interesting and profitable work – not as a consultant, as he imagined at first, but as a contract worker. There is a renewed interest in woodlands, brought about in part by the need to find other uses for farmland, and Mike is well-placed to take

advantage of it. Similarly, farmers are shedding labour as the recession bites and making use of more contractors and casual workers to carry out specific jobs. Again, Mike is in a good position to take advantage of the situation.

Betty was lucky to recapture the milk round from Borchester Dairies who purchased it when Mike got into difficulties; they now have it on a franchise and pay the dairy 1p a pint for the privilege. It fits in well with Mike's life-style in as much as it demands early rising, something to which a cowman and dairy farmer is well-accustomed. When he first took the milk round he received some advice from a man who had been dropping pints on people's doorsteps for years. 'Forget all this stuff about "your friendly neighbourhood milkman"' he was told. 'Get out there as early as you can and make your deliveries before most of them are up. That way you don't have to waste time listening to Mrs Bloggs prattling on about her cat. You'll get through in half the time and be home before breakfast.' Having a rural round Mike doesn't quite manage that but he is usually through by 9.30 and ready for the other half of his job.

Mike Tucker may surprise all of us one of these days.

Neil Carter came to Brookfield in the spring of 1973 as a 'new entrant', the current vogue term for an apprentice. Although he had no previous farming experience he settled quickly to the life and when he reached the end of his training period, opted to stay with the Archers. He developed a liking for the pigs and, while able to do more or less anything on the farm, spent much of his time at Hollowtree. He grew into a sincere, hardworking, hard to ruffle, conscientious young man, but inside the gentle and modest farmworker there beat the heart of a budding entrepreneur. By 1981 he was talking of starting his own pig enterprise while continuing to work at Brookfield, but Phil was afraid of his bringing disease from his own pigs to Hollowtree and the idea was dropped. The principle survived, however, and the following year saw him start up in egg production in one of the poultry houses at Ambridge Farm. Before long he had 1000 birds in battery cages, seeing to them before and after work and building up a profitable egg round in the Borchester area. It was all accomplished quietly with remarkably little fuss, and only stopped when Mike Tucker moved in to Ambridge Farm and told Neil he would soon be needing the poultry house for calves.

Neil's plans floundered for a while until Bill Insley, the retired Derbyshire farmer who had bought Willow Farmhouse and 15 acres, came to the rescue with the offer of a barn to house the battery cages. Neil and the avuncular Mr Insley built up an affinity for each other and it wasn't long before there was talk of a share-farming agreement. Share farming is an arrangement whereby one party provides the land, buildings, and fixed equipment, and the other provides labour, the mobile machinery and expertise. Both parties provide part of the working capital and each receives part of the end-product in return for his investment. Bill Insley's idea was that Neil could run a herd of outdoor pigs on his land and fatten the porkers in his buildings. This arrangement suited Neil perfectly since his lack of capital could be balanced by Bill's ample funds and compensated for in the final splitting of the finished porkers. Bill was confident that Neil knew his pigs and in April 1986 they started with the first of 30 gilts. Neil negotiated a contract with Phil Archer to carry on running the Hollowtree pig unit by promising to buy his gilts from the same source, thus minimising the risks of disease.

Over the summer they gradually built up the gilts at the rate of six each month, mating them on arrival to stagger the farrowings and so make do with the minimum of buildings. Everything was going according to plan with the first litter due in August and the first porkers ready just before Christmas when suddenly Bill Insley died of a heart attack. It was only 6 months since the arrangement started and before a pig was sold. Neil's future looked bleak until it was revealed that the benevolent Mr Insley had left him the barn (which housed his hens and fattening pigs) and the 8 acres on which he was keeping the sows. Under the terms of the will Neil had to repay his executors £5000 over the next seven years for Bill's share of the livestock and fixed equipment, a sum his benefactor must have calculated as being just about possible from the budgeted profits.

Since then Neil has savoured both the satisfaction and the pressures of running a small business. He has experienced the thrill of putting his sows to the boar, of seeing them farrow and of loading their progeny on to a lorry bound for the abattoir. He has also suffered the humiliation of being sent for by the bank manager, when the pig price slumped, and told to reduce his overdraft. However, the greatest lesson he has learned is how to work flat out month after month for up to 14 hours a day.

A 'normal' day for Neil begins before 7 o'clock when he calls in briefly at Willow Farm to check that everything is as it should be before going on to see to the pigs at Hollowtree. He usually finishes there soon after 10 o'clock and then goes back to tend his own stock. He's due at Hollowtree again at about 4 pm to give the afternoon feed and then returns to do the same at Willow Farm. But superimposed on what sounds like a reasonably relaxed working day are a host of crises and additional demands on Neil's time. Sows don't always farrow conveniently while he's feeding either at Hollowtree or Willow Farm and he's constantly having to dash from one place to the other to

supervise. Sometimes a difficult farrowing keeps him up half the night. An initial check on his own hens and pigs may reveal trouble in the form of a sick pig, or a frozen pipe or a broken fence which has to be dealt with straight away, delaying his start at Hollowtree and making him late for the rest of the day. Frequently he has to work extra time with the Archers' pigs – when they are bringing in new gilts, or weighing or sending off baconers. Quite often he has undertaken to work elsewhere for several hours during the day to bring in extra cash. Two or three evenings a week he's out doing his egg round to earn the essential premium over packing station prices, and most of his never-ending backlog of maintenance work at Willow Farm has to be carried out at weekends. It's not surprising that Susan complains that she and the children hardly ever see him.

When he and Bill Insley first planned the pig enterprise they spoke of starting with 30 sows and building up to 100. At the moment it's doubtful if Neil could finance expansion even if he wanted to. But the tragic experience of Mike Tucker together with 18 months of disastrous pig prices in 1987 and 1988 have curbed his ambitions somewhat, though with his present activities bringing in £10 000 or £11 000 a year he cannot be too despondent. Now aged 31, he has probably realised that he's unlikely to make it as a full-blown farmer; on the other hand he won't willingly go back to full-time employment. Neil is the sort of chap who proceeds pragmatically but he may well settle for a permanent blend of self-employed-farmworker-cum-smallholder.

Those who knew Shula Archer in her early teens would in all likelihood have predicted a 'horsey' future for her. Everything pointed in that direction: her flair for riding, the hours spent at her Aunt Chris's stables and the horse management course she attended for three terms. She had passed her Pony Club 'A' test, visited Ann Moore's riding establishment and was all set to compete at Hickstead and Wembley. Then she started working as a junior clerk at Rodway and Watson's, the Borchester firm of auctioneers, estate agents and surveyors, and enjoyed herself so much that she decided to stay. Eight years later, now Mrs Hebden, she became a full-blown Associate of the Royal Institution of Chartered Surveyors.

Rodway's is a typical country firm embracing a wide spectrum of activities from selling properties to running Borchester livestock market, carrying out valuations and managing estates. Although Shula had experience of all these operations during her years of qualification, she has recently been more occupied in running the Bellamy estate. She often gives the impression that she is in the driving seat, but in fact defers regularly to Mr Rodway. After all, he is the one who makes the final decision as to whether Joe Grundy's rent will be reduced, and he is the one who signs the cheques. This has been necessary partly due to Shula's lack of experience and partly to anticipate any possible clash of interest resulting from her dealing with relatives. There's no doubt that she acquired this particular responsibility through nepotism and it has been difficult for her trying to reconcile the interests of two cousins: the owner of the estate Lilian Bellamy, and Tony Archer, one of its tenants. With Aunt Peggy working in the estate office, her brother David applying for the post of estate farm manager and her cousin-by-marriage Brian Aldridge doing the odd spot of contract work, there have been times when Shula has been only too glad to lean on Mr Rodway and rely on him for the last word.

Shula spends about three days a week in the estate office dealing with everything from checking rate demands to deciding whether the Girl Guides can camp at Valley Farm. Budgets and cash flows have to be prepared, insurances reviewed, forestry grant schemes sorted out. The farm manager works direct to Mr Rodway but Shula must liaise with him regularly over matters where farm and estate responsibilities overlap, such as felling trees on the farms. Mr Rodway comes once a week to go through things with her and sign cheques. The rest of the week she spends in Borchester on other Rodway and Watson business, such as preparing sales plans and repair specifications.

The Bellamy estate, which Shula manages, is about 1000 acres; it's the residue retained by Lilian when the estate was broken up in 1975 on the death of Ralph Bellamy. It comprises seven farms, three of which are let and four 'in hand' (farmed by the estate):

LAND IN HAND

Sawyer's Farm } Heydon Farm } run as one unit (mainly dairying)	260 acres
Valley Farm (rearing young stock, forage)	160 acres
Ambridge Farm (all arable, milk quota transferred to other farms)	150 acres
Woodland	110 acres

TENANTED FARMS

Grange Farm (Joe Grundy)	120 acres
Bridge Farm (Tony Archer)	140 acres
Red House Farm (Leonard Roberts)	80 acres

Over the years Shula has become a key figure in the Ambridge scene as well as a valuable member of the staff of Rodway and Watson with whom she has recently become an associate. Her long-term future as a land agent will, no doubt, depend on whether or not she starts a family.

'We'll just have to wait and see'

'Thirsty work, talking', quips Brian as he puts two pints of Shires on the polished table. He and Phil have met by arrangement at The Bull to discuss the idea of including Brookfield land in the Home Farm shoot, but the chat has now moved on to future plans for both their holdings.

'The things is we don't know what the scientists will come up with next,' says Phil, taking a swig of his beer. 'You were talking about BST just now but that's not the only bit of biotechnology in the pipeline is it? I hear they've got a similar drug for pigs on the way – makes 'em produce meat with less fat on less feeding stuffs. That would help us up at Hollowtree.'

'Trouble is every other producer would have it as well – so who'd benefit in the long run? Not the poor old pig, that's for sure. Joe Public, I s'pose, getting his pork at half the price.' Brian leans back. 'Then he'd complain it gave him some ghastly disease or other.'

'This work they're doing on getting cereals to fix their own nitrogen sounds interesting – you know, like legumes. Wonder whether we'll see the fruits of that before the end of the century.'

'Something's got to happen before then or we'll all be

broke,' replies Brian. 'I was going through my figures with the accountant this week and our returns are way down again.'

Phil sips his beer. 'It's the cows that have been our salvation these last few years. I hate to think where we'd be without the milk cheque, especially now that David's married and we've got two families to keep on the place. Heaven knows what we'll do if Kenton wants a share of the action.'

'Well, here's someone who doesn't have to worry about the future,' laughs Brian as Tony Archer comes into the bar. 'Come over here and cheer us up. We're wondering whether we'll still be in business by the year 2000.'

'It's nearer than you think,' says Tony, pulling up a chair as Brian goes to fetch him a drink. 'Pat and I were only talking about it last night. Do you know that if I had one of my cows inseminated next New Year's Day, and she brought me a heifer calf, I'd still be milking it in the twenty-first century?'

'It's a funny old world,' Brian observes as he sits down, 'the only one of us whose profit's going up each year is Tony. Isn't that right?'

Tony grins. 'That's 'cos I'm giving the public what they want. Cheers!'

'But how long do you see the demand for your organic stuff continuing?' asks Phil.

'It'll only pay as long as you get a good mark-up on it compared with Phil's and mine!' Brian interjects.

'I reckon the premium for organic produce is here to stay,' Tony replies. 'On the one hand you've got all this worry about health and food additives which is going to sustain demand or even increase it. And on the other you've got all the hard work involved in producing it which is going to put a lot of farmers off.' He takes a quick drink. 'I think there'll always be people prepared to pay more for food without chemicals and I reckon that there aren't going to be all that many folk prepared to face the slog. It's nearly put Pat and me off a time or two.'

'So you'll press on with the yoghurt and stuff?' Phil demands.

'For the time being, anyway. It sells for ten times as much as I'd get for the milk. What concerns me is that if it goes on growing I shall end up as a factory manager, worrying about staff and bad debts and the tax man rather than real farming.'

'Still, better than having to scrape around for the rent,' says Phil, toying with the beer mat. 'Talking about factories, Brian, do you reckon there's any future in producing stuff for industrial use – you know, like extracting ethanol from your beet instead of sugar?'

Brian screws up his face. 'There's all sorts of things they *could* make from what we grow, from motor fuel to plastics. Trouble is, to make it viable, either the cost of oil has to shoot up or the price we get for our stuff has to come down. One's not very likely and I don't like the sound of the other. I s'pose I could put up a plant to extract starch from your potatoes, Phil. How do you think that would go down?'

'Mrs Snell would be out with the petitions before you could say Maris Piper, I should think. Remember what happened when you were thinking of putting up the pig unit?'

'And I reckon there'll be a lot more Lynda Snells in Ambridge before you get round to putting up your starch plant,' Tony comments. 'That bloke I rent the field off in the village is itching to get permission to cover it with houses. I'm surprised he hasn't got it already.'

'Be a good thing in some ways,' says Phil. 'Keep village life going – you know bolster the WI, help fill the church, keep Martha's shop open. We can't put a glass case over Ambridge, and farming certainly isn't going to produce that many jobs in the future.'

Tony chuckles. 'Might even rescue the cricket club from the bottom of the league.'

'It'll just make our job more difficult,' Brian snorts. 'They'll be traipsing around all the footpaths, complaining about the noise of the grain dryer or the bird scarers, or when we burn a bit of straw.'

Phil couldn't help joining in. '*And* the smell of slurry-spreading *and* the hedge trimmings . . .'

'P'raps we can cash in on them,' says Tony. 'You know, open a farm shop, flog 'em stuff at the back door. Do cream teas, charge 'em to watch us do the milking, that sort of thing.'

'It's probably the only way we *will* survive in the end,' says Brian, with feeling, 'turning our farms into leisure centres – providing fishing, boating, shooting, riding. There's a lot of money to be made if it's well done. But do we really want to get *involved*? We're farmers at heart.'

Phil decides to change the subject. 'I think a lot depends on what happens in politics and marketing, both here and in Brussels. I mean what's going to happen to the CAP? And what will 1992 mean to us farmers?'

'It means Tony will be selling his yoghurt in Lyons,' Brian cuts in.

'Or the French'll be supplying milk to London – through the Chunnel,' says Tony. 'Anyway, what's going to happen to quotas? *And* to the Milk Board for that matter. Is that going to survive?'

'And the Potato Board,' Phil adds. 'If that goes I doubt if I'll be growing any more spuds.'

'I might though. It'll be a large-scale operation then,' says Brian, putting down his glass. 'But seriously, where are we all going to be by 2000 AD? What shall we all be doing, I wonder?'

They all stare at their empty tankards. Tony is the first to reply, a little hesitatingly. 'Well, as I said earlier, it's not all that far off. We could have a nice family business going by then at Bridge Farm, p'raps turning all our milk into dairy products of one sort or another, butter, cheese and ice-cream as well as yoghurt. Even buying in organic milk. There'd be jobs for all the kids. That's if we can stand the pace.'

Brian speaks next. 'The most *sensible* thing for me to do is to sell up and live on the proceeds. I'd be very rich, even after the tax man had a go. But where would I live? And what would I do? First thing tycoons do when they make a few million is to buy a farm, isn't it? Or p'raps I should put the whole of Home Farm down to set-aside and sack all the chaps.' He eyes his glass. 'But I won't. I may have a closer look at this leisure business. As I said, I think there's money to be made there.'

Now it's Phil's turn. 'I think we'll still be in business. Mind you, I'll be turned 70 by then so I shan't want to be racing round much. We'll probably increase the cows, lease some quota if that's still going, and double the sheep. Concentrate on quality lamb *and* beef, probably using this new embryo transfer technique for getting beef calves from dairy cows. It's possible that David might get fed up waiting for me to retire and take another farm in the meantime – but they'll come back. There've been Archers at Brookfield for 150 years and I reckon they'll still be there next century. 'Course, we can't be sure can we, any of us? We'll just have to wait and see.'

'Well,' says Tony, gathering up the glasses, 'as I'm the only one who seems to be making any money at the moment, it must be my shout. Who's for another?'

Index

Figures in *italics* refer to illustrations